ROBOTICS AND AUTOMATION IN HEALTHCARE

Advanced Applications

Frontiers of Mechanical and Industrial Engineering

ROBOTICS AND AUTOMATION IN HEALTHCARE

Advanced Applications

Edited by

R. Thanigaivelan, PhD

Sanjay Singh, PhD

Clement Christy Deepak C., PhD

AAP | APPLE ACADEMIC PRESS

First edition published 2025

Apple Academic Press Inc.
1265 Goldenrod Circle, NE,
Palm Bay, FL 32905 USA

760 Laurentian Drive, Unit 19,
Burlington, ON L7N 0A4, CANADA

CRC Press
2385 NW Executive Center Drive,
Suite 320, Boca Raton FL 33431

4 Park Square, Milton Park,
Abingdon, Oxon, OX14 4RN UK

Library and Archives Canada Cataloguing in Publication

Title: Robotics and automation in healthcare : advanced applications / edited by R. Thanigaivelan, PhD, Sanjay Singh, PhD, Clement Christy Deepak C., PhD.
Names: Thanigaivelan, Rajasekaran, 1977- editor. | Singh, Sanjay (Dean of the School of Engineering and Technology at Dhanalakshmi Srinivasan University), editor. | Deepak, C. Clement Christy, editor.
Description: First edition. | Series statement: Frontiers of mechanical and industrial engineering |Includes bibliographical references and index.
Identifiers: Canadiana (print) 20240307178 | Canadiana (ebook) 20240307259 | ISBN 9781774915486 (hardcover) | ISBN 9781774915493 (softcover) | ISBN 9781003424161 (ebook)
Subjects: LCSH: Robotics in medicine. | LCSH: Medical care—Automation. | LCSH: Artificial intelligence—Medical applications. | LCSH: Medical technology.
Classification: LCC R857.R63 R63 2024 | DDC 610.285/63—dc23

Library of Congress Cataloging-in-Publication Data

CIP data on file with US Library of Congress

ISBN: 978-1-77491-548-6 (hbk)
ISBN: 978-1-77491-549-3 (pbk)
ISBN: 978-1-00342-416-1 (ebk)

ABOUT THE FRONTIERS OF MECHANICAL AND INDUSTRIAL ENGINEERING BOOK SERIES

EDITORS-IN-CHIEF

Dr. Prasenjit Chatterjee
Department of Mechanical Engineering, MCKV Institute of Engineering, West Bengal, India

Dr. Dilbagh Panchal
Department of Industrial and Production Engineering, Dr. B. R. Ambedkar National Institute of Technology (NIT) Jalandhar, Punjab, India

Description of the Series

Mechanical engineering applies the principles of mechanics and materials science for analysis, design, manufacturing, and maintenance of mechanical systems, whereas industrial engineering involves figuring out how to make or do things better with minimum resource consumption. Mechanical and industrial engineering utilize a deep foundational understanding of mathematics, physics, and analysis to develop machines and systems. The former specializes in thermodynamics, combustion, and electricity to make complex machines. The latter focuses on designing workflow and making production efficient.

This series reports on the cutting-edge developments to present novel innovations in the field of mechanical and industrial engineering. Volumes published in this series aim to embrace all aspects, subfields, and new challenges of mechanical and industrial engineering and to be of interest and use to all those concerned with research in various related domains of mechanical and industrial engineering disciplines. The series aims to provide comprehensive value-added resources to the body of the literature to serve as essential references for years ahead. Features of the volumes under this series explore recent trends, model extensions, developments, solutions to real-time problems, and case studies.

The coverage of the series is committed to provide a diverse yet rigorous research experience that builds on a solid knowledge of science and engineering fundamentals and application-oriented knowledge of mechanical and industrial engineering practices for efficient functioning in today's global multidisciplinary rapidly changing work environment. It includes all new theoretical and experimental findings in mechanical, industrial engineering, and allied engineering fields.

Topics and Themes:

The primary endeavor of this series is to introduce and explore contemporary research developments in a variety of rapidly growing research areas. The volumes will deal with the following topics but not limited to:

- Automotive engineering
- Aerospace technology
- Big data in mechanical / manufacturing / industrial engineering
- Computational methods for optimizing manufacturing technology, robotics, mechatronics
- Decision science applications in mechanical, manufacturing, and industrial engineering
- Design and optimization of mechanical components
- Dynamical systems, control
- Engineering design
- Engineering thermodynamics, heat, and mass transfer
- Evolutionary computation
- Finite element method (FEM) modeling/simulation
- Fluid mechanics
- Fuzzy logic and neuro-fuzzy systems for relevant engineering applications
- Green and sustainable engineering techniques for modern manufacturing
- Innovative approaches for modeling relevant applications and real-life problems
- Life cycle engineering
- Machinery and machine elements
- Manufacturing
- Mathematical concepts and applications in mechanical and industrial engineering
- Materials engineering
- Mechanical structures and stress analysis
- MEMS
- Modeling in engineering applications
- Nanotechnology and microengineering
- Non-conventional machining in modern manufacturing systems

- Precision engineering, instrumentation and measurement
- Numerical simulations
- Soft computing techniques
- Sustainability in mechanical, manufacturing, and industrial engineering
- Theoretical and applied mechanics
- Tribology and surface technology

The series aims to serve as a valuable resource for undergraduate, postgraduate, doctoral students, researchers, academicians, and industry professionals in mechanical, manufacturing, industrial engineering, product design, management, and more.

BOOKS IN THE SERIES

Optimization Methods for Engineering Problems
Editors: Dilbagh Panchal, PhD, Prasenjit Chatterjee, PhD,
Mohit Tyagi, PhD, and Ravi Pratap Singh, PhD

Modern Manufacturing Systems: Trends and Developments
Editors: Rajiv Kumar Garg, PhD, Ravi Pratap Singh, PhD,
Rajeev Trehan, PhD, and Ramesh Singh, PhD

Multi-Criteria Decision-Making Methods in Manufacturing Environments: Models and Applications
Shankar Chakraborty, PhD, Prasenjit Chatterjee, PhD, and
Partha Protim Das, PhD

Supply Chain Performance Measurement in Textile Enterprises
Editors: Pranav G. Charkha, PhD, Santosh B. Jaju, PhD, and
Prasenjit Chatterjee, PhD

Electro-Micromachining and Microfabrication: Principles and Research Advances
Editors: Sandip Kunar PhD, Golam Kibria, PhD, and Prasenjit Chatterjee PhD

Robotics and Automation in Healthcare: Advanced Applications
Editors: R. Thanigaivelan, PhD, Sanjay Singh, PhD, and
Clement Christy Deepak C., PhD

Optimization of Advanced Manufacturing Processes
Editors: Sandip Kunar, PhD, Prasenjit Chatterjee, PhD, and
M. Sreenivasa Reddy, PhD

ABOUT THE EDITORS

R. Thanigaivelan, PhD
Principal, A.K.T. Memorial College of Engineering and Technology, Kallakurichi, Tamil Nadu, India

R. Thanigaivelan, PhD, is working as a Principal at A.K.T. Memorial College of Engineering and Technology, Kallakurichi, Tamil Nadu, India. He is a graduate of Mechanical Engineering from Madras University in 1999 and an MTech in Mechatronics from Vellore Institute of Technology, Vellore in 2004. He obtained his PhD from Anna University, Coimbatore, in 2012. He has 18 years of teaching experience and 13 years of research experience. He has published over 75 papers in SCI and SCOPUS indexed journals and has supervised 11 PhD scholars. He has edited books from IGI Global and authored a chapter in a book composed by CRC Press, Taylor and Francis, USA. He is a member of the doctoral candidates for three scholars. He has acted as a reviewer for various SCI journals. His areas of interest are micromachining, surface engineering, composites, and biomaterials. He is a member of the Institution of Engineers (India), Quality Circle Forum of India (QCFI), India Society of Technical Education (ISTE), and Indian Institute of Production Engineers (IIPE).

Sanjay Singh, PhD
Dean–School of Engineering and Technology, Dhanalakshmi Srinivasan University, Tiruchirapalli, Tamil Nadu, India

Sanjay Singh, PhD, is presently working as Dean of the School of Engineering and Technology at Dhanalakshmi Srinivasan University, Tamil Nadu, India. He has 18 years of working experience in the aviation industry and has contributed towards the upkeep of aircraft and aero engines. He also did research work in the diagnostic maintenance of internal combustion engines using the Spectrometric Oil Analysis Program (SOAP), utilizing the equipment called the Atomic Absorption (AA) Spectrometer.

He was in a leadership position for training ab-initio trainees and young boys and girls in schools and colleges for preparing them for the second line of defense force in the National Cadet Corps (NCC). He switched over from industry to academic and has 13 years of teaching experience with additional administrative responsibilities as Head of the Department. He authored 15 papers in SCI- and SCOPUS-indexed journals. His areas of interest are tribology, non-destructive testing, composites, preventive, and predictive maintenance, and robotics. He is a lifetime member of the Institution of Engineers, India, and the Society for Shock Wave Research, created by the Indian Institute of Science, Bengaluru, Karnataka, India. Dr. Singh is a graduate of Mechanical Engineering (the Institution of Engineers, India) and earned a master's degree in Automation and Robotics and a PhD at Vinayaka Missions Research Foundation, Deemed to be University, Salem, Tamil Nadu, India.

Clement Christy Deepak C., PhD
Manager, Gears and Gearbox Projects,
Eppinger Tooling Asia Pvt Ltd, Kondampatti,
Coimbatore, Tamil Nadu, India

Clement Christy Deepak C., PhD, is a Manager of Gears and Gearbox projects at Eppinger Tooling Asia Pvt Ltd, Kondampatti, Coimbatore, Tamil Nadu, India. He completed a Bachelor's in Mechanical Engineering in 2010, a Master's in Engineering CAD/CAM in 2013, a Master of Business Administration in 2016, and a Doctorate in Philosophy in Mechanical Engineering in 2021. He boasts a teaching experience of six years and industrial experience of five years. He published papers in various national and international journals. He published two patents and conducted many funded seminars and workshops for the students' development. He is an editorial board member and reviewer for national and international journals. He published books on "Manufacturing Technology-I" and "A Concept AWD Mechanism in Automobile Applications". He has guided many projects for UG students and received the Best Project Award in 2014. He is a member of many professional bodies.

CONTENTS

Contributors .... *xiii*

Abbreviations .... *xvii*

Preface .... *xxi*

1. **Wireless Charging of Implantable Pacemaker: Design, Analysis, and Comparison of GaN Rectifier Circuits 1**
   Punithavathi Duraiswamy

2. **Building Smart Bot Designs for Library Assistance 21**
   S. Valai Ganesh, V. Suresh, S. Godwin Barnabas, and S. Rajakarunakaran

3. **Evaluating the Biomechanical Effects of Titanium Dental Implants of Different Sizes on Strain Distribution Using the Strain Gauge Method 45**
   Duraisamy Velmurugan, Thangavel Yuvaraj, and Rajendran Selvapriya

4. **Depth Estimation Using Stereo Medical Imaging 61**
   Ketki P. Kshirsagar and Rajeshree A. Shinde (Rokade)

5. **Development of Drones for Healthcare Applications 75**
   S. A. Menon Rajkumar, S. Kalaibharath, T. Niranjan, A. Gokul, and R. Thanigaivelan

6. **Modeling and Optimization of 3D Printing Processes: A Literature Review 95**
   Jyotisman Borah and M. Chandrasekaran

7. **Applications of Medical Image Processing and Computer Vision in Healthcare 115**
   S. Sudhagar, S. Suresh, P. M. Gopal, and S. Seenivasan

8. **Actuators in Medical Devices 137**
   S. Sasikumar, S. Santhakumar, R. Jayapal, and R. Thanigaivelan

9. **Smart Spectacles with Embedded Seizure Detector 151**
   M. Soundarrajan, J. Preetha, M. Aravindh, and M. Vishnu

10. **A TOPSIS Method for Industrial Service Robot Selection with Different Weight Approach 169**
    D. Deepa, S. Suresh, K. G. Saravanan, and Raghavendra Datta Naik

11. Aerial Robotics: An Overview of the Medical Applications of Drone Systems 187
C. Ramprasadh

12. Advanced Mechanical Ventilator 207
U. Nithishkumar, S. Sendil Kumar, J. Meghala, and J. Sundaraperumal

13. Development of a Medical Decision Support System for Early Detection of Cardiac Disorders 223
S. Sumathi and S. Ponlatha

14. Robots and Cyborgs with Artificial Intelligence-Based Technologies in the Healthcare Sector: A Review 239
S. Ponlatha and S. Sumathi

15. Biomedical and Healthcare Applications of Nanorobots 257
P. Suresh, U. Saravanakumar, S. Senthilkumar, M. Suganthi, T. Aravind, and V. K. Buvanesvari

Index 279

CONTRIBUTORS

T. Aravind
Department of CSE, Muthayammal Engineering College, Rasipuram, Namakkal, Tamil Nadu, India

M. Aravindh
Muthayammal Engineering College, Rasipuram, Namakkal, Tamil Nadu, India

S. Godwin Barnabas
Department of Mechanical Engineering, Ramco Institute of Technology, Rajapalayam, Virudhunagar, Tamil Nadu, India

Jyotisman Borah
Research Scholar, Department of Mechanical Engineering, NERIST, Arunachal Pradesh, India

V. K. Buvanesvari
Department of MCA, Muthayammal Engineering College, Rasipuram, Namakkal, Tamil Nadu, India

M. Chandrasekaran
Professor, Department of Mechanical Engineering, NERIST, Arunachal Pradesh, India

D. Deepa
Department of Mechanical Engineering, Muthayammal Engineering College, Rasipuram, Namakkal, Tamil Nadu, India

Punithavathi Duraiswamy
Department of Electronics and Communication Engineering, M.S. Ramaiah University of Applied Sciences, Bangalore, Karnataka, India

S. Valai Ganesh
Department of Mechanical Engineering, Ramco Institute of Technology, Rajapalayam, Virudhunagar, Tamil Nadu, India

A. Gokul
Department of Mechanical Engineering, Muthayammal Engineering College, Rasipuram, Namakkal, Tamil Nadu, India

P. M. Gopal
Center for Materials Science, Mechanical Engineering, Karpagam Academy of Higher Education, Coimbatore, Tamil Nadu, India

R. Jayapal
Department of Mechanical Engineering, Muthayammal Engineering College, Rasipuram, Namakkal, Tamil Nadu, India

S. Kalaibharath
Department of Mechanical Engineering, Muthayammal Engineering College, Rasipuram, Namakkal, Tamil Nadu, India

Ketki P. Kshirsagar
Electronics and Tele-Communication Department, Vishwakarma Institute of Information Technology, Pune, Maharashtra, India

S. Sendil Kumar
Department of Electrical and Electronics, S. A. Engineering College – Autonomous, Chennai, Tamil Nadu, India

J. Meghala
Department of Electrical and Electronics, S. A. Engineering College – Autonomous, Chennai, Tamil Nadu, India

Raghavendra Datta Naik
Department of Mechanical Engineering, Goa College of Engineering, Goa, India

T. Niranjan
Department of Mechanical Engineering, Mahatma Gandhi Institute of Technology, Hyderabad, Telangana, India

U. Nithishkumar
Department of Electrical and Electronics, S. A. Engineering College–Autonomous, Chennai, Tamil Nadu, India

S. Ponlatha
Department of Electronics and Communication Engineering, Mahendra Engineering College, Namakkal, Tamil Nadu, India

J. Preetha
Muthayammal Engineering College, Rasipuram, Namakkal, Tamil Nadu, India

S. Rajakarunakaran
Department of Mechanical Engineering, Ramco Institute of Technology, Rajapalayam, Virudhunagar, Tamil Nadu, India

S. A. Menon Rajkumar
Department of Mechanical Engineering, Muthayammal Engineering College, Rasipuram, Namakkal, Tamil Nadu, India

C. Ramprasadh
School of Mechanical Engineering, SASTRA Deemed to be University, Tirumalaisamudram, Thanjavur, Tamil Nadu, India

S. Santhakumar
Department of Mechanical Engineering, Muthayammal Engineering College, Rasipuram, Namakkal, Tamil Nadu, India

U. Saravanakumar
Department of ECE, Muthayammal Engineering College, Rasipuram, Namakkal, Tamil Nadu, India

K. G. Saravanan
Department of Mechanical Engineering, Sona College of Technology, Salem, Tamil Nadu, India

S. Sasikumar
Department of Mechanical Engineering, Muthayammal Engineering College, Rasipuram, Namakkal, Tamil Nadu, India

S. Seenivasan
Department of Mechanical Engineering, Dhaanish Ahmed Institute of Technology, Coimbatore, Tamil Nadu, India

Rajendran Selvapriya
Department of Civil Engineering, Muthayammal Engineering College, Rasipuram, Namakkal, Tamil Nadu, India

S. Senthilkumar
Department of EIE, VNE Vignana Jyothi Institute of Engineering and Technology, Hyderabad, Telangana, India

Rajeshree A. Shinde (Rokade)
Electronics and Tele-Communication Department, Lokmanya Tilak College of Engineering, Navi Mumbai, Maharashtra, India

M. Soundarrajan
Muthayammal Engineering College, Rasipuram, Namakkal, Tamil Nadu, India

S. Sudhagar
Department of Mechanical Engineering, Muthayammal Engineering College, Namakkal, Tamil Nadu, India

M. Suganthi
Department of ECE, Mahendra College of Engineering, Salem, Tamil Nadu, India

S. Sumathi
Department of Electrical and Electronics Engineering, Mahendra Engineering College, Namakkal, Tamil Nadu, India

J. Sundaraperumal
Department of Electrical and Electronics, S. A. Engineering College–Autonomous, Chennai, Tamil Nadu, India

P. Suresh
Department of Mechanical Engineering, Muthayammal Engineering College, Rasipuram, Namakkal, Tamil Nadu, India

S. Suresh
Department of Mechanical Engineering, Erode Sengunthar Engineering College, Erode, Tamilnadu, India

V. Suresh
Department of ECE, National Engineering College, Kovilpatti, Thoothukudi, Tamil Nadu, India

R. Thanigaivelan
Department of Mechanical Engineering, Muthayammal Engineering College, Rasipuram, Namakkal, Tamil Nadu, India

Duraisamy Velmurugan
Department of Mechanical Engineering, Muthayammal Engineering College, Rasipuram, Namakkal, Tamil Nadu, India

M. Vishnu
Muthayammal Engineering College, Rasipuram, Namakkal, Tamil Nadu, India

Thangavel Yuvaraj
Department of Mechanical Engineering, Muthayammal Engineering College, Rasipuram, Namakkal, Tamil Nadu, India

ABBREVIATIONS

3D	three-dimensional
A3C	asynchronous actor-critic algorithm
ABS	acrylonitrile butadiene styrene
AED	automated external defibrillator
AHP	analytical hierarchy process
AI	artificial intelligence
AMD	autonomous mental development
ANN	artificial neural network
AOA	angle of attack
AVE	average variance extracted
$BaTiO_3$	barium titanate
BCI	brain-computer interface
BiPAP	bilevel positive airway pressure
BTBS	brain-to-brain system
CB-SEM	covariance-based SEM
CCA	common carotid artery
CCD	central composite design
CG	computer-generated
CI	cochlear implant
CIS	computerized information systems
CPAP	continuous positive airway pressure
CsI	cesium iodide
CT	computed tomography
CVD	cardiovascular disease
DBS	deep brain stimulation
DC	direct current
DEA	data envelopment analysis
DES	dielectric elastomer substrate
DFD	depth from the defocus
DMLS	direct metal laser sintering
DOF	degrees of freedom
DPG	deep deterministic policy gradient
DQN	deep Q network
DSD	double-screw-drive

ECG	electrocardiogram
EEG	electroencephalography
ER	evolutionary robotics
ESC	electronic speed controller
FAA	Federal Aviation Administration
FAD	fuzzy axiomatic design
FDM	fused deposition modeling
FEA	finite element analysis
FFF	fused filament fabrication
FIT	field innovation team
FL	fuzzy logic
FPD	flat panel detector
FPV	first-person view
GA	genetic algorithm
GaAs	gallium arsenide
GaN	gallium nitride
GPS	global positioning system
GTRI	Georgia Tech Research Institute
HAI	hospital-acquired infection
HIV	human immunodeficiency virus
HTMT	heterotrait-monotrait ratio
IC	indigo carmine
ICNP	incorporated carbon nanotube position
ICU	intensive care units
IoT	internet of things
ISM	industrial, scientific, and medical
ISTAR	intelligence, surveillance, target acquisition, and reconnaissance
LiDAR	light detection and varying
LiPo	lithium polymer
LRS	limit reduction strategy
Mask R-CNN	mask-R convolutional neural networks
MCDM	multi-criteria decision-making
MI	myocardial infraction
MIS	minimally invasive surgery
ML	machine learning
MRF	Markov random field
MRI	magnetic resonance imaging
NCC	normalized cross-correlation

NEMS	nanoelectromechanical systems
NHLS	national health laboratory service
NMP	neuromotor prostheses
NSR	normal sinus rhythm
OCT	optical coherence tomography
OLS	ordinary least squares
PD	proportional-derivative
PEEK	polyether ether ketone
PET	positron emission tomography
PETG	polyethylene terephthalate glycol
PIS	positive ideal solution
PLA	polylactic acid
PLS-SEM	partial least squares SEM
PP	positive predictivity
PSO	particle swarm optimization
PVC	premature ventricular contraction
RF	radio frequency
SAD	sum of absolute difference
SCs	speed controllers
Se	sensitivity
SEM	structural equation modeling
SEM-SE	scanning electron microscopy–secondary electron
SGBM	semi-global block-matching
SGM	semi-global simulation
SiC	silicon carbide
SL	stereo lithography
SLM	selective laser melting
SMA	shape memory alloys
SPECT	single-photon emission computed tomography
SSVEP	steady-state visual evoked potential
STAR	smart tissue autonomous robot
STFT	short-term Fourier transform
STL	stereolithography
TCBs	transfer connection blocks
TFT	thin-film transistor
TOPSIS	technique for order of preference by similarity to ideal solution
UAV	unmanned aerial vehicles
UTM	universal testing machine

VF	ventricular fibrillation
VFLU	ventricular flutter
VLOS	visual line-of-sight
VNS	vagus nerve stimulation
VT	ventricular tachycardia
VTOL	vertical take-off and landing
WHO	World Health Organization
WPT	wireless power transmission
WPU	waterborne polyurethane
WT	wavelet transform

PREFACE

This book addresses the various issues related to robotics and automation in healthcare. The recent pandemic has forced researchers to adopt technologies such as robotics and AI in the healthcare field. This book discusses the various aspects of robotics and automation, including robot selection and development, drones, image processing, and identification.

Chapter 1 discusses the various aspects of wireless charging of implantable pacemakers. Chapter 2 focuses on smart bot design for the library building of a medical college. Chapter 3 discusses the strain distribution in biomechanical systems. The subsequent chapters address the issues in medical imaging, drone technology, 3D printing, and image processing techniques.

The application and importance of actuators in medical devices, especially during surgery is also discussed. Other topics the book covers include wearable devices for pre-identification of seizure development in patients, decision support system for the detection of suitable robots, and early detection of diseases with the support of image processing techniques. The application of nanorobots in healthcare is also observed in the last chapter. Hence, this book gives insight into robotics, wearable devices, and the application of image processing in the healthcare field.

CHAPTER 1

WIRELESS CHARGING OF IMPLANTABLE PACEMAKER: DESIGN, ANALYSIS, AND COMPARISON OF GAN RECTIFIER CIRCUITS

PUNITHAVATHI DURAISWAMY

Department of Electronics and Communication Engineering, M.S. Ramaiah University of Applied Sciences, Bangalore, Karnataka, India

ABSTRACT

A new power transfer method for a variety of uses is called wireless power transmission, or WPT. In the medical field, a small pacemaker with a remote-chargeable battery is indispensable. High power handling WPT systems are needed to meet future demand for a range of applications. GaN Schottky diodes can function at high frequencies and with high power. It can be used to remotely charge the pacemaker's batteries because of its high-power tolerance. In order to increase the wireless power transmission and energy harvesting efficiency of the WPT system, the rectifier is essential. The impedance matching networks designed using micro strips with optimum size helps to maximize the efficiency by proving high return loss. The comparison of simulation results show that the voltage doubler rectifier gives high efficiency for wide range of input power with high resistance loads. The performance of the GaN rectifier circuits has to be investigated with an input power less than 0 dBm as the received signal in practical cases will be weak. By demonstrating a large return loss, the impedance matching networks created with micro strips of optimal size contribute to optimizing

Robotics and Automation in Healthcare: Advanced Applications. R. Thanigaivelan, Sanjay Singh, & Clement Christy Deepak C. (Eds.)

efficiency. The voltage doubler rectifier provides good efficiency for a wide range of input power with high resistance loads, according to a comparison of simulation findings. In practical scenarios, the received signal will likely be faint, hence it is necessary to evaluate the performance of the GaN rectifier circuits at input powers lower than 0 dBm.

1.1 INTRODUCTION TO WIRELESS POWER TRANSMISSION (WPT)

In this time of modernization, electrical power plays a vital role. Wireless power or remote power means transmission of electrical power from a power source to an electrical load without interconnect wires. Currently, a huge number of power-constrained remote communication systems are used for various applications ranging from transportable entertainment systems to healthcare devices. These systems are commonly controlled by batteries that have a restricted process time and they have to be replaced or re-energized in order to assurance the process of a network for a specific time frame. However, the strategy of battery substitution or recharging may build an undesired use and, at times, is not practical, e.g., implanting gadgets such as pacemakers inside human bodies.

Pacemakers are needed for people who have heart rhythmic disorders. They are a tiny battery-operated device that stimulates the heart and makes the heartbeat at correct pace. Sometimes, patients after a heart attack or heart surgery need a temporary pacemaker. Temporary pacemakers are connected to a power source using wires, and it causes complications. Some of the complications include infection from the wires crossing the skin, scar tissue formation, and damage to the heart upon removal. Alternatively, pacemakers can be powered wirelessly, using WPT technology. It aims to eliminate the battery burdens by charging the batteries from a distance.

WPT has been emerging as a promising technology and are being developed at various operating frequencies for different applications [1–3]. Though research and demonstrations have been carried out at different frequencies, 2.4 GHz emerged as the transmitting frequency of choice as it lies in the Industrial, Scientific, and Medical (ISM) band. The key component, rectenna (antenna + rectifier) in the WPT system determines the converter efficiency [4]. Several rectifier circuit topologies are available with single diodes or dual diodes.

The WPT system shown in Figure 1.1 consists of transmitter and receiver. On the transmitter, there is an oscillator that generates Radio Frequency (RF) signal, followed by a power amplifier and transmitting antenna. The main components of receiver are receiving antenna and rectifier. Together they are called rectenna [5].

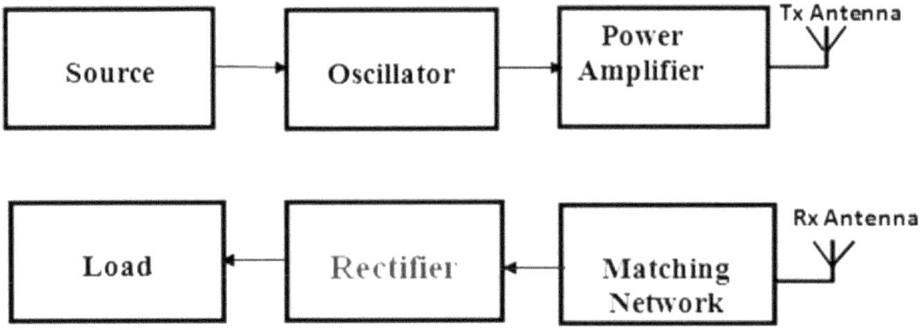

FIGURE 1.1 Block diagram of wireless power transmission system.

In WPT system, rectifier converts RF signal captured from antenna to Direct Current (DC) and it crucial for obtaining high efficiency. There are different types of rectifiers available in the literature. Based on material used, they are classified as Silicon rectifier, Germanium rectifier or compound material such as gallium nitride rectifier, etc. Based on the topology, popularly classified rectifiers are series, shunt, and voltage doubler rectifier.

Rectifier efficiency plays an important role in transferring the charge to the load. Another main element of WPT framework is matching network that is responsible for lossless transmission between source and load.

1.1.1 RECTIFIER EFFICIENCY AND MATCHING NETWORK

The efficiency (η) of rectifier is defined as the ratio of DC rectified output to RF input signal. Efficiency can be increased with the use of low voltage drop diode and smoothing circuits which reduce the ripple content of output. Another important circuit is the matching circuit responsible for lossless transmission between the source and its load, i.e., from antenna to the rectifier at the chosen operating frequency of 2.4 GHz [6, 7]. RF-to-DC conversion efficiency is calculated using Eqn. (1) through Eqn. (3):

$$\eta = \frac{P_{dc}}{P_{RF}} \times 100 \quad (1)$$

$$P_{dc} = \frac{V_{dc}^2}{RL} \quad (2)$$

$$P_{RF} = Vrms \times Irms \tag{3}$$

where; R_L is the load resistor.

1.1.2 SILICON DIODE RECTIFIER

Diodes are used for rectification, and Schottky diodes are preferred in WPT applications. Silicon Schotkey diode rectifiers using various topologies such as series, shunt, voltage doubler is available in literature. However, switching speed of silicon diode is slower than the silicon carbide (SiC), gallium arsenide (GaAs) and gallium nitride (GaN) semiconductor diodes and have very low voltage overshoot and silicon diode cannot handle high power, high-frequency RF applications.

1.1.3 GAN DIODE RECTIFIER

Research studies shows that wide band gap GaN devices are better than Silicon devices. GaN Schottky diodes are expected to provide higher power handling capability because of wide bandgap [8–11]. In modern years, GaN has risen as an extremely capable material for both electronic and optoelectronic strategy generally due to its broadband gap (3.4 eV), and most appropriate for high power applications [8]. Unlike a conventional PN-junction diode, which is formed by combining P-type and N-type semiconductor material, Schottky Diodes are constructed by joining a metal electrode and N-type semiconductor. Therefore, they are significantly faster as they are unipolar devices. Their speed is only limited by the junction capacitance. The switching time is in the range of picoseconds for the small-signal diodes and up to tens of nanoseconds for special high-power diodes. Figure 1.2 shows the 2009 to 2015 market forecast for SiC and GaN device industry in power electronics.

1.2 DESIGN OF GAN DIODE RECTIFIERS

The methodology to design the impedance matching circuit in order to get high efficiency is depicted in Figure 1.3. Typical matching impedance used is 50 Ω and the operating frequency is 2.4 GHz. A capacitive load of 7.8 pf is used. This impedance matching circuit and the rectifier design is carried

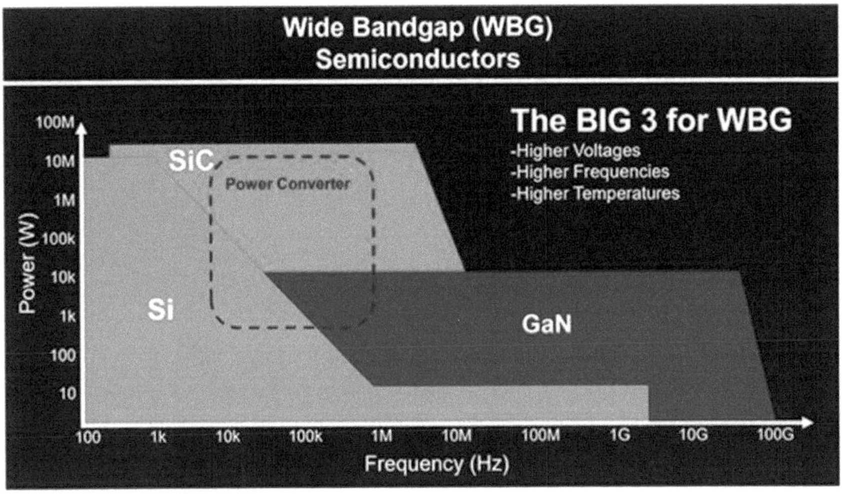

FIGURE 1.2 GaN and SiC market projection.

Source: www.kemet.com.

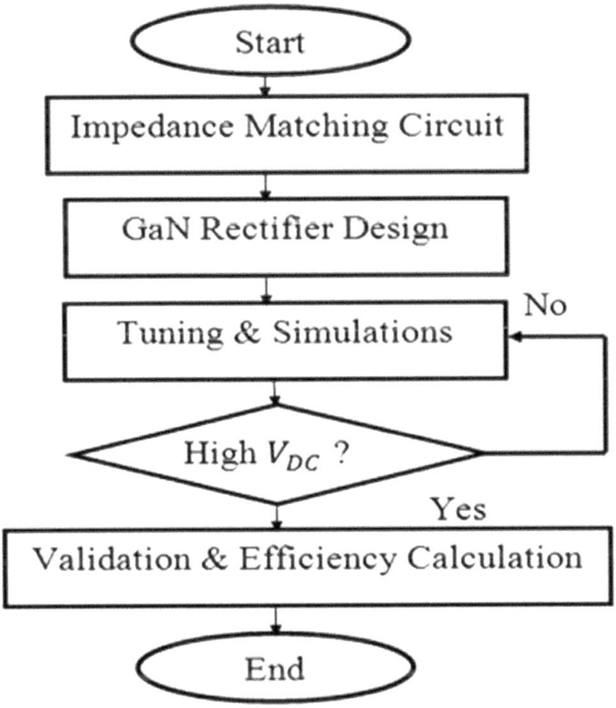

FIGURE 1.3 Methodology for designing microstrip matching circuit.

out using ADS 2008 EDA tool. A microstrip matching circuit is used in this design. The microstrip length and width is varied and the S-parameter simulations are carried out. The output voltage and the efficiency are calculated by giving an input. The procedure is repeated and tuning the length and width of the microstrip is done to obtain maximum efficiency with optimum microstrip dimensions.

1.2.1 GAN DIODE CIRCUIT MODEL

Models of the Silicon Schottky diode model and GaN Schottky diode circuit are reported in literature [11, 12]. The GaN Schottky diode model proposed in Ref. [11] is used in this rectifier design. The various parameters of GaN Schottky diode used in this design are listed in Table 1.1. Three different rectifiers are designed and simulated in this design. Series, shunt, and voltage doubler rectifier circuits are designed, analyzed, and compared with efficiency as main objective.

TABLE 1.1 The Electrical Parameters of GaN Schottky Diode [11]

Name	Parameter	Value
Is	Reverse saturation current	3 e–13 mA
Rs	Series resistance	7.2 Ω
N	Ideality factor	1.5
Cjo	0 V junction capacitance	0.19 pF
Vj	Built-in voltage	1.0 V
M	Capacitance grading coefficient	0.28

1.3 GAN RECTIFIER TOPOLOGIES

The three microwave rectifier circuit topologies, series diode rectifier, shunt diode rectifier, and voltage doubler rectifier with microstrip matching network, are shown in Figure 1.4. GaN Schottky barrier diode is used for rectification because of its low forward voltage and high-frequency switching capability compared to Silicon PN diode or Silicon Schottky diode. The performance comparison of the three topologies is studied and analyzed.

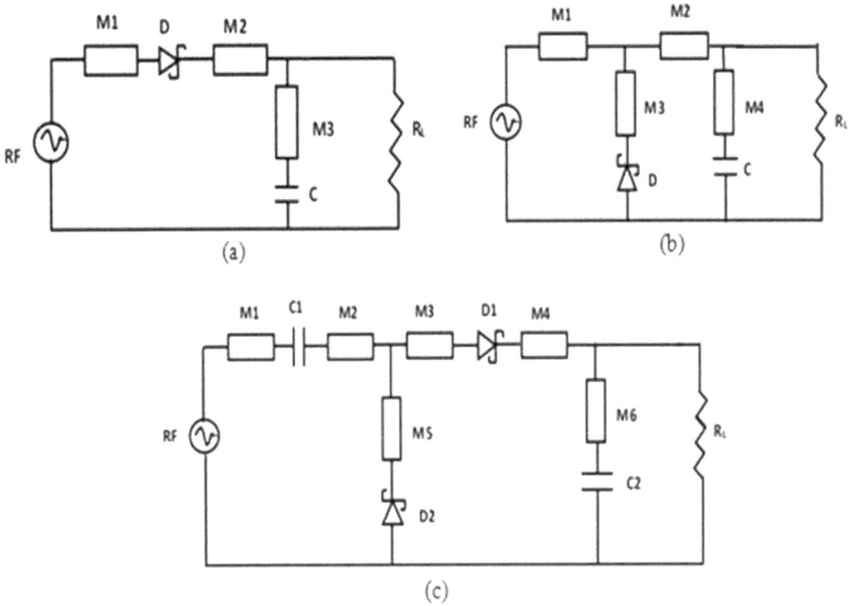

FIGURE 1.4 (a) series diode rectifier; (b) shunt diode rectifier; and (c) voltage doubler rectifier [13].

1.4 IMPEDANCE MATCHING NETWORK

1.4.1 SERIES HALF-WAVE GAN RECTIFIER

The design of impendence matching circuit is shown in Figure 1.5 for series half wave rectifier. Impendence matching network uses microstrip line matched to R_L of 50 Ω at frequency of 2.4 GHz. Using tuning option in the ADS tool microstrip line width and length are changed for a good return loss (S11). The impedance-matching network consists of microstrip lines M1, M2, and M3. Two microstrips M1 and M2 are inserted before and after the diode improves the DC output voltage and thereby the converter efficiency. M3 helps in shifting the center frequency. The S-parameter simulation shows S11 of 37 dB @ 2.4 GHz as indicated as shown in Figure 1.6(a) at 2.4 GHz frequency. S11 values are measured in dB and Smith chart is used to measure the impedance is shown in Figure 1.6(b). The -10 dB bandwidth starts at 2.35 GHz and end at 2.46 GHz, which is about 100 MHz.

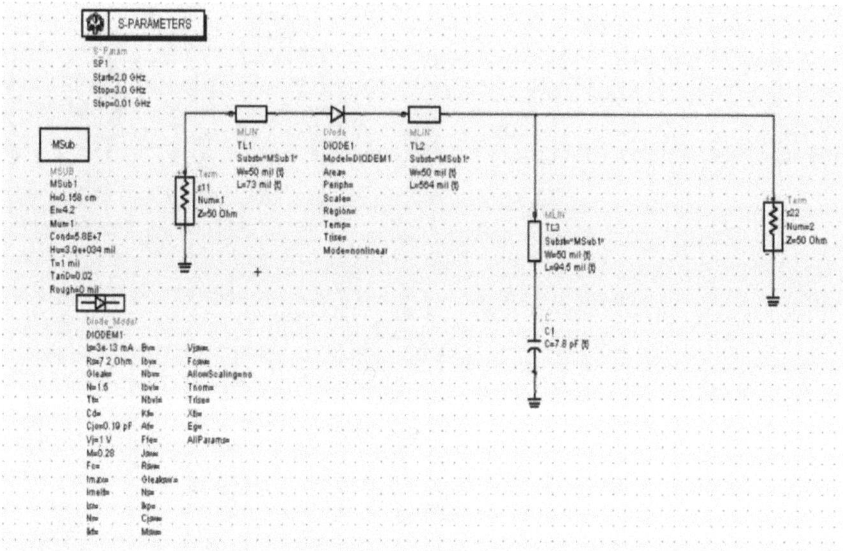

FIGURE 1.5 Series half-wave impedance matching circuit.

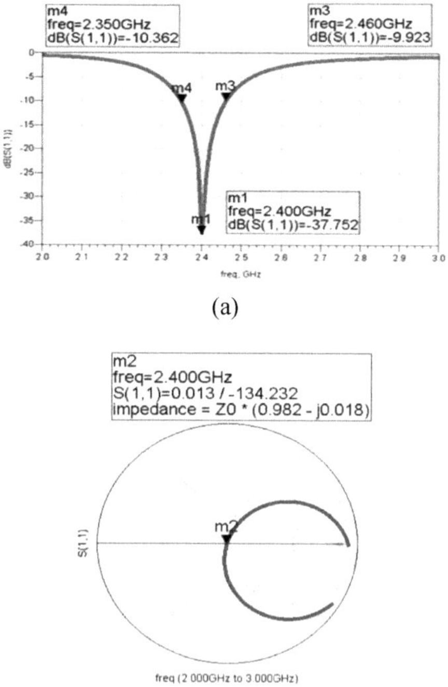

(a)

(b)

FIGURE 1.6 (a) Return loss S11; and (b) Smith chart.

1.4.2 SHUNT HALF WAVE GAN RECTIFIER

Figure 1.7 shows the shunt half-wave rectifier topology using a single GaN diode matched to 50 Ω resistance. The microstrips are included both in series and shunt with the components. A return loss of 57 dB is obtained using optimized microstrip dimensions. M1, M2, and M3 are used to improve the rectifier efficiency and M4 is used to match the circuit to 2.4 GHz. It is observed that for both series diode and shunt diode circuits, the micro-strips inserted close to the capacitor is longer than the rest of the microstrips.

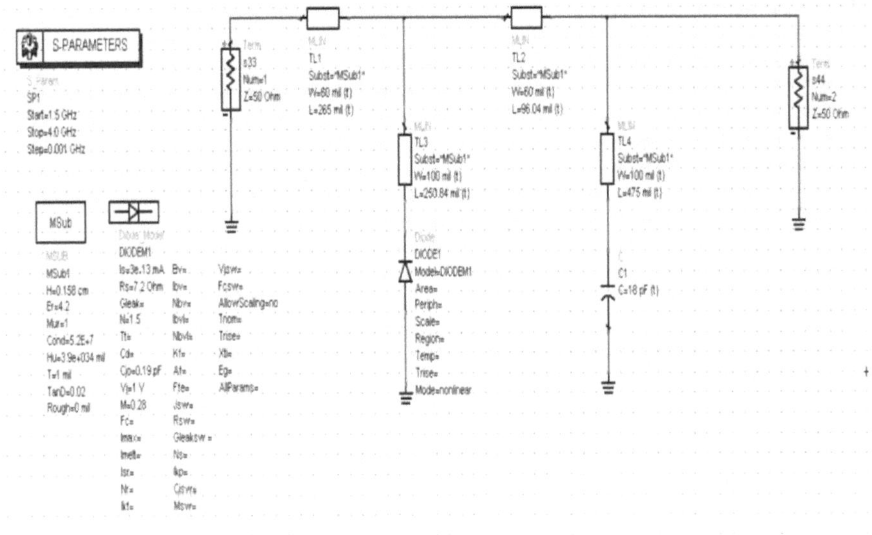

FIGURE 1.7 Shunt half-wave impedance matching circuit.

Figure 1.8(a) shows the shunt GaN rectifier return loss (S11) and Smith chart is shown in Figure 1.8(b) at 2.4 GHz frequency. The -10 dB bandwidth starts at 2.06 GHz and end at 3.1 GHz, which is about 1.06 GHz.

1.4.3 VOLTAGE DOUBLER GAN RECTIFIER

The impendence matching circuit for voltage doubler rectifier circuit is show in Figure 1.9.

The microstrips inserted before and after the diodes (M1, M2, M3, and M4) increases the efficiency of the circuit. The R_L is kept close to the capacitor (C2). Unlike the series and shunt diode rectifiers, it is seen that the microstrips connected in the shunt arm is small compared to the other microstrips. A return loss of 64 dB is obtained. The -10 dB bandwidth starts at 2.2 GHz and end at 2.57 GHz, which is about 310 MHz. Figure 1.10 shows the return loss and Smith chart.

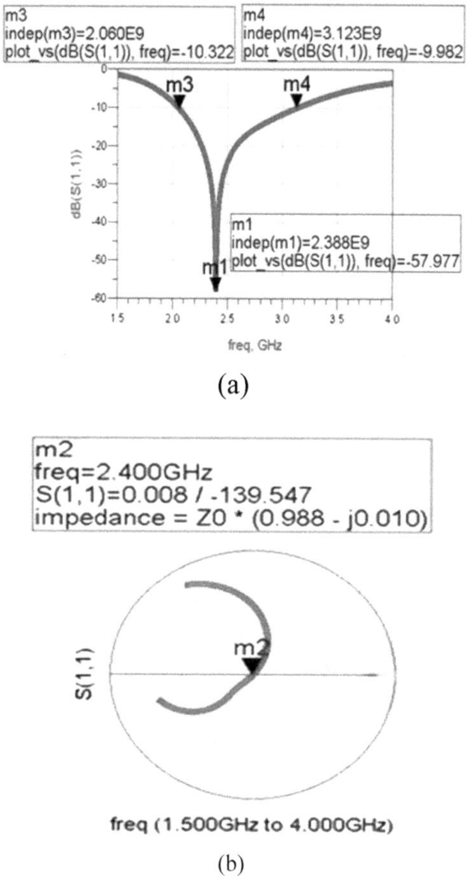

FIGURE 1.8 (a) Return loss (S11); and (b) Smith chart.

1.5 OPTIMIZED MICROSTRIP DIMENSIONS

The optimized microstrip dimensions designed for the three rectifiers are shown in Table 1.2. These dimensions are used, and the transient simulations are carried out to measure the output voltage and calculate the efficiency.

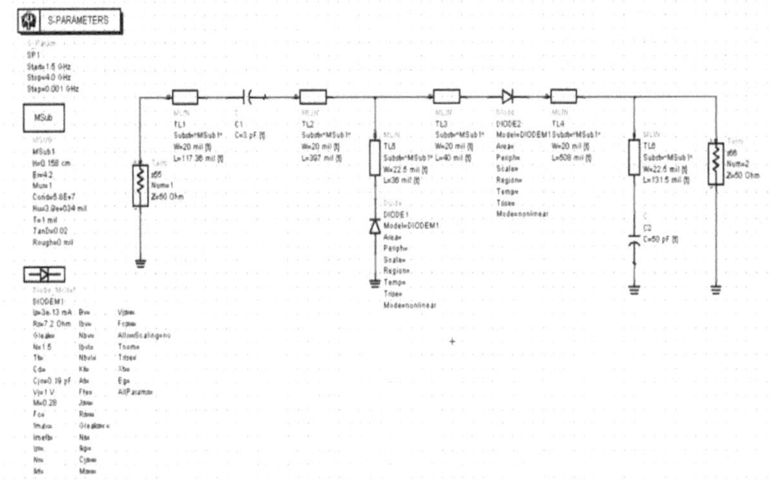

FIGURE 1.9 Voltage doubler impedance matching circuit.

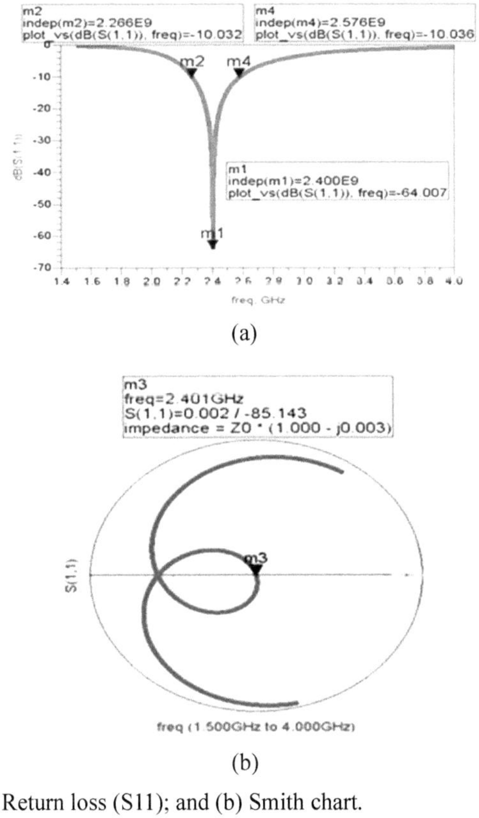

(a)

(b)

FIGURE 1.10 (a) Return loss (S11); and (b) Smith chart.

TABLE 1.2 Optimized Microstrip Dimension [13]

	M1	M2	M3	M4	M5	M6
Series diode rectifier	1.8 mm × 1.5 mm	14 mm × 1.2 mm	23.5 mm × 1.25 mm	–	–	–
Shunt diode rectifier	6.6 mm × 1.5 mm	2.4 mm × 1.5 mm	6.25 mm × 2.5 mm	11 mm × 2.5 mm	–	–
Voltage doubler rectifier	2.9 mm × 0.5 mm	9.9 mm × 0.5 mm	1 mm × 0.5 mm	12 mm × 0.5 mm	1 mm × 0.56 mm	3.2 mm × 0.56 mm

1.6 SIMULATION RESULTS

1.6.1 SERIES HALF-WAVE GAN RECTIFIER

The series diode half wave rectifier circuit designed for operating at 2.4 GHz is shown in Figure 1.11. Two microstrips M1 and M2 are inserted before and after the diode. It improves the DC Output voltage and thereby the converter efficiency. M3 helps in shifting the operating frequency. Besides, it is also found that when a microstrip is inserted between the capacitor and the load, the DC output voltage drops, and the ripples increase. With an input voltage (V_{in}) of 5 V, the output voltage (V_{dc}) obtained is 2.32 V with a R_L of 150 Ω. This gives a conversion efficiency of 69%. Transient simulation of input and output results are shown in Figure 1.12.

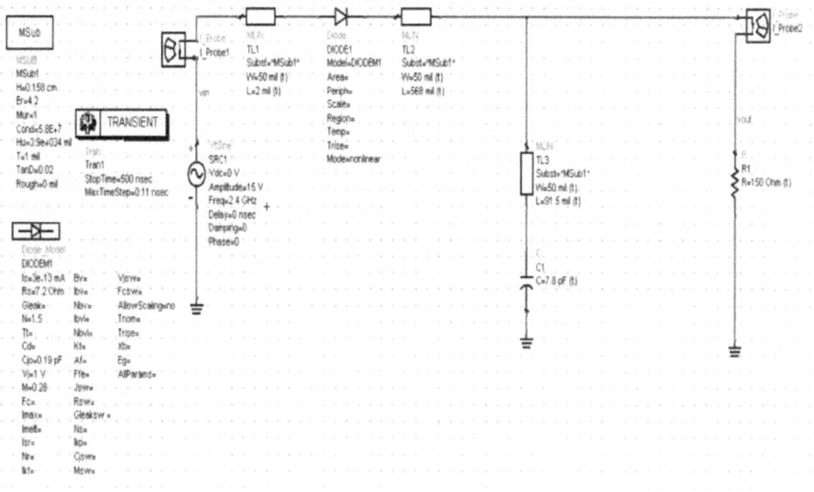

FIGURE 1.11 Design of series GaN rectifier at 2.4 GHz.

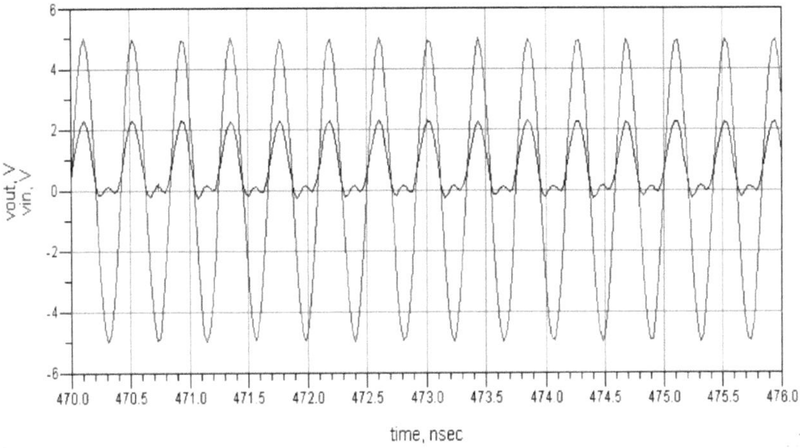

FIGURE 1.12 Simulation results of series GaN rectifier at 2.4 GHz.

1.6.2 SHUNT HALF-WAVE GAN RECTIFIER

Transient simulations are carried out for the half-wave shunt rectifier shown in Figure 1.13. The output voltage is measured, and the efficiency is calculated. Figure 1.14 shows an output voltage (V_{dc}) of 9.88 V with an input voltage (V_{in}) 15 V with R_L 60 Ω. This is equal to an efficiency of 63.6%.

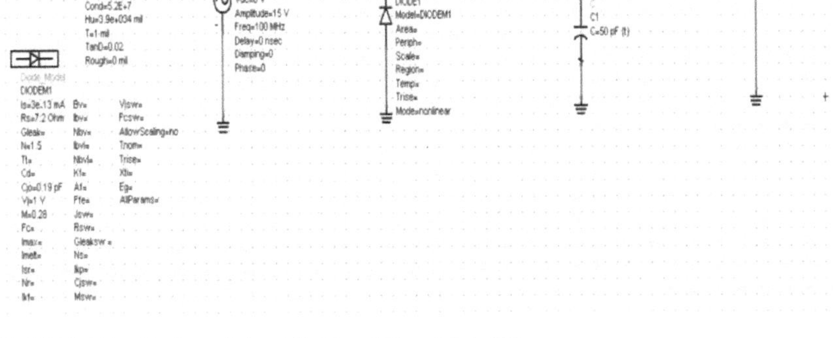

FIGURE 1.13 Design of shunt GaN rectifier at 2.4 GHz.

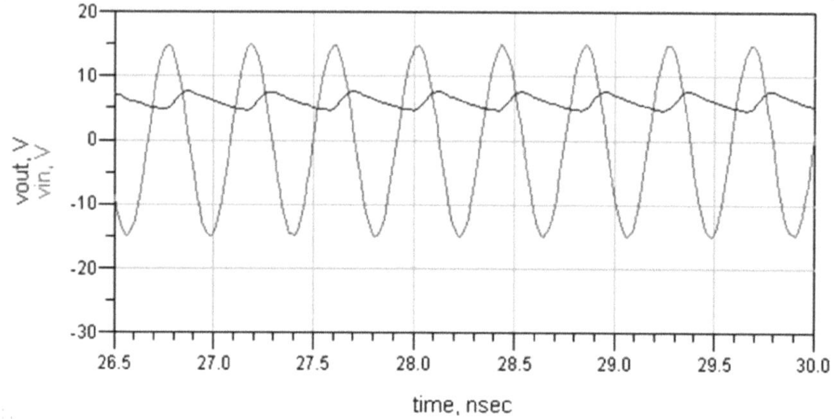

FIGURE 1.14 Simulation results of shunt GaN rectifier at 2.4 GHz.

1.6.3 VOLTAGE DOUBLER GAN RECTIFIER

Figure 1.15 shows a Villard's voltage doubler. It is created by adding two arrangements of half-wave rectifiers in a cascading arrangement, in this topology dual GaN Schottky diode and dual capacitors, each pair rectifies half cycle.

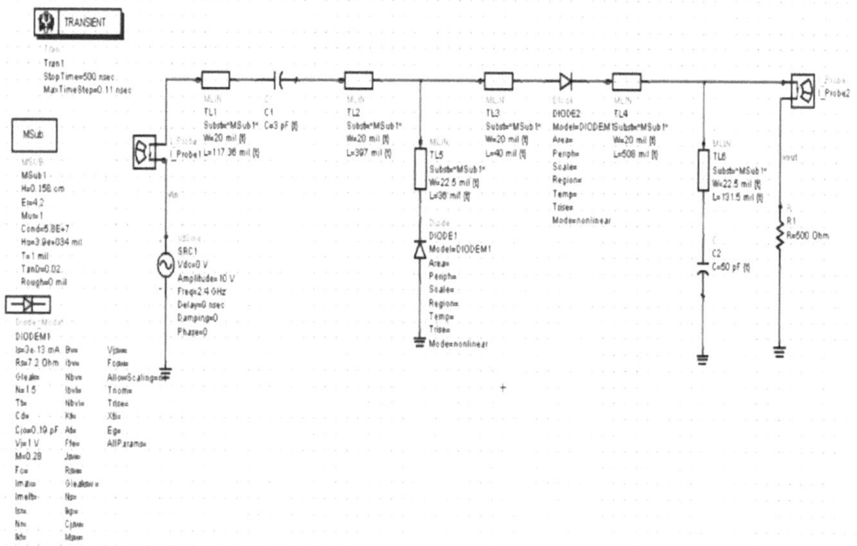

FIGURE 1.15 Design of Villard's voltage doubler GaN rectifier at 2.4 GHz.

For the period of negative half cycle, diode D1 is forward biased, starts conducting and the capacitor C1 charges to input voltage. At same instance, diode D2 is reverse biased. During the positive half cycle of the input, diode D2 is forward biased and capacitor C2 starts charging. This time capacitor charges to twice of the input. The efficiency obtained is 78.2% with a load resistance of 1,000 Ω. The transient simulation waveforms for the input and output are shown in Figure 1.16.

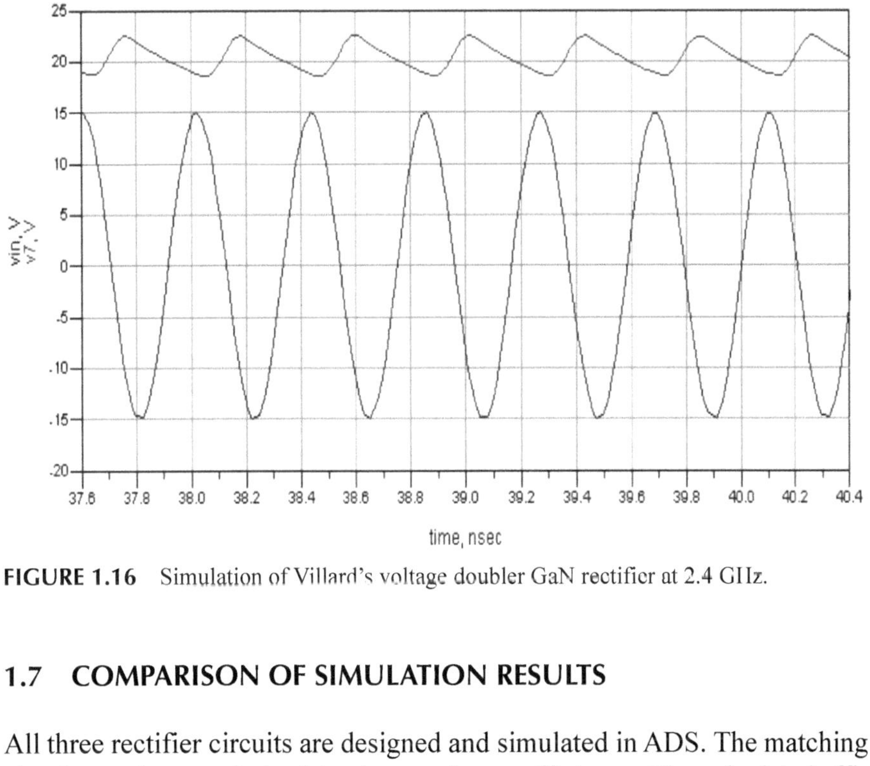

FIGURE 1.16 Simulation of Villard's voltage doubler GaN rectifier at 2.4 GHz.

1.7 COMPARISON OF SIMULATION RESULTS

All three rectifier circuits are designed and simulated in ADS. The matching circuits used are optimized to give maximum efficiency. The calculated efficiency is compared, and its dependency on load resistance, input power, and frequency are analyzed as shown in Figures 1.17–1.19, respectively. From the simulation results of Figure 1.17, it is observed that voltage doubler provides an efficiency of more than 60% for R_L greater than 500 Ω. It is also seen for low resistive loads. The efficiencies of series and shunt diode rectifiers are high. For the input power dependency simulation in Figure 1.18, a minimum input power of 15 dBm is maintained. It is observed that series diode and voltage doubler rectifier circuits maintain high efficiency for a wide range of input power. However, the shunt diode rectifier requires high input power to have a good converter efficiency compared to series diode and voltage doubler rectifier circuits. The series diode rectifier is able to maintain a converter efficiency of over 50% and the voltage doubler rectifier maintains an efficiency of 65%. The frequency dependent efficiency plot in Figure 1.19 shows high efficiency of 60% for all three rectifiers in the frequency range of 2.4 GHz–2.5 GHz. Table 1.3 gives the efficiency of the simulated rectifiers with the input power of 34 dBm. It is clear that the voltage doubler rectifier efficiency performance is better than the single diode rectifiers. The efficiency obtained for voltage doubler rectifier is 78.2% with R_L of 1 KΩ.

TABLE 1.3 Comparison between Three GaN Rectifiers Based on Efficiency [13]

Rectifier Configurations	Frequency (GHz)	V_{in}	Load (Ω)	η (%)
Series diode half wave rectifier	2.4 GHz	15	150	69
Shunt diode half wave rectifier	2.4 GHz	15	60	63.6
Voltage doubler	2.4 GHz	15	1,000	78.2

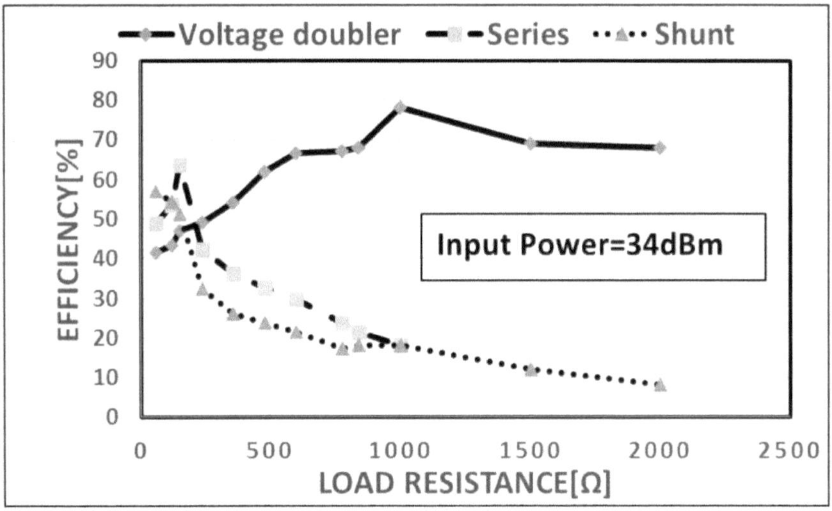

FIGURE 1.17 Comparison of load resistance dependency simulation [13].

Analysis using various loads, input power, and efficiency at various frequencies This section explains the characteristics and reveals the performance of three typical microwave series, shunt half-wave rectifiers, and voltage doubler rectifiers by using the same GaN Schottky diode. Comparison results show that the voltage doubler diode rectifier performance is better than the other two rectifiers. Voltage doubler GaN rectifier topology is well suited for remote transmission of power for various applications.

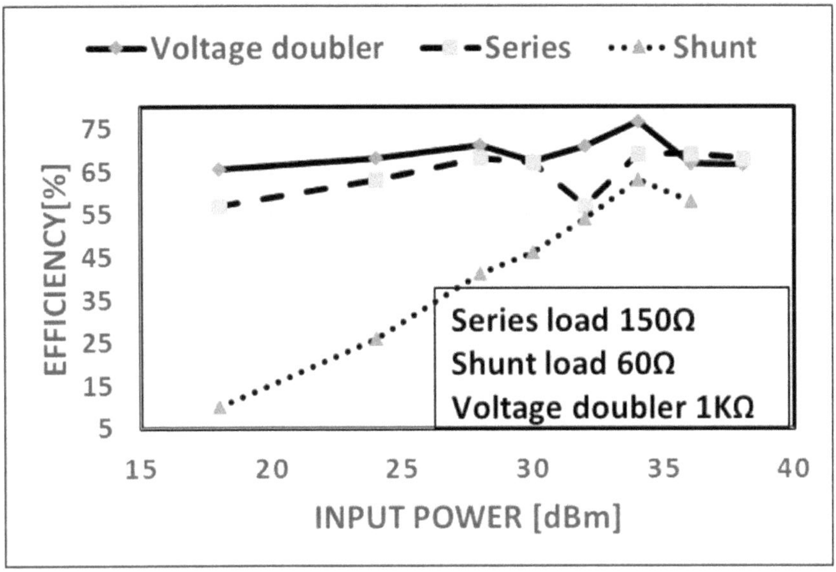

FIGURE 1.18 Comparison of input power dependency simulation [13].

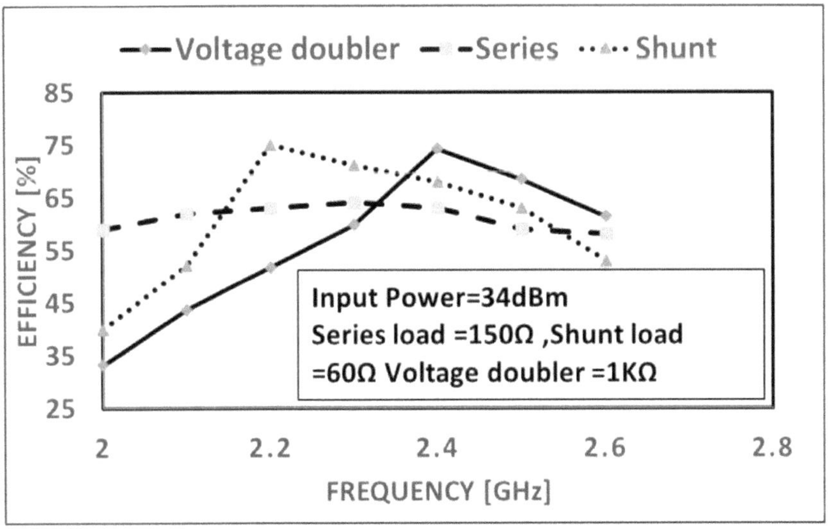

FIGURE 1.19 Comparison of frequency dependency simulation [13].

1.8 CONCLUSION

WPT is emerging as a power technology for the transfer of power for various applications. The need for a tiny pacemaker whose battery can be charged from a distance is essential in the field of healthcare. A WPT system that can handle high power is required to meet the growing demands for various applications in the future. GaN Schottky diodes can withstand high power and can be operated at high frequencies. As it can withstand high power, it can be used to charge the batteries of pacemakers from a distance. The rectifier plays a key role in improving the efficiency of the WPT system in WPT and energy harvesting. The impedance matching networks designed using microstrips with optimum size help maximize efficiency by proving high return loss. The comparison of simulation results shows that the voltage doubler rectifier gives high efficiency for a wide range of input power with high resistance loads. The performance of the GaN rectifier circuits has to be investigated with an input power of less than 0 dBm as the received signal will be weak in practical cases.

KEYWORDS

- **direct current**
- **gallium arsenide**
- **radio frequency**
- **return loss**
- **silicon carbide**
- **wireless power transmission**

REFERENCES

1. Karalis, A., et al., (2008). Efficient wireless non-radiative midrange energy transfer. *Annals of Physics*, *323*, 34–48.
2. Ladan, S., & Wu, K., (2014). 35 GHz harmonic harvesting rectifier for wireless power transmission. *Microwave Symposium (IMS), 2014 IEEE MTT-S International* (pp. 1, 4).
3. Fang, Z., Fan-Yi, M., Jong-Chul, L., & Qun, W., (2013). Study of a novel compact rectenna for wireless energy harvesting. *Wireless Symposium (IWS), 2013 IEEE International* (pp. 1–4).

4. Omar, A., & Campana, E., (2010). *Study of the Efficiency of Rectifying Antenna Systems for Electromagnetic Energy Harvesting.* Thesis, Universitat Politècnica de Catalunya.
5. Wang, D., & Negra, R., (2012). Design of a rectifier for 2.45 GHz wireless power transmission. *PhD Research in Microelectronics and Electronics (PRIME), 2012 8th Conference* (pp. 1–4). Aachen, Germany.
6. Doan, H. C., & Bach, G. D., (2015). Investigation of rectifier circuit configurations for microwave power transmission system operating at S-band. *International Journal of Electrical and Computer Engineering (IJECE), 5*(5), 967–974.
7. Berndie, S., & Chang, K., (2003). 5.8-GHz circularly polarized dual rhombic-loop traveling-wave rectifying antenna for low power density wireless power transmission applications. *IEEE Transactions on Microwave Theory and Techniques, 51*, 1548–1553.
8. Millán, J., & Godignon, P., (2013). Wide bandgap power semiconductor devices. In: *2013 Spanish Conference on Electron Devices* (pp. 293–296). Valladolid.
9. Millán, J., (2012). A review of WBG power semiconductor devices. *CAS 2012 (International Semiconductor Conference)* (pp. 57–66). Sinaia.
10. Jain, H., Rajawat, S., & Agrawal, P., (2008). Comparison of wide bandgap semiconductors for power electronics applications. *Recent Advances in Microwave Theory and Applications, 2008: Microwave 2008; International Conference, Jaipur* (pp. 878–881).
11. Efthymiou, L., Longobardi, G., Camuso, G., Hsieh, A. P. S., & Udrea, F., (2015). Modeling of an AlGaN/GaN Schottky diode and extraction of main parameters. *Semiconductor Conference (CAS), 2015 International* (pp. 211–214). Sinaia.
12. Takacs, A., Bunea, A., Zair, A., Neculoiu, D., & Aubert, H., (2012). Diode modeling for millimeter wave applications based on co-simulation technique. *CAS 2012 (International Semiconductor Conference)* (pp. 201–204). Sinaia.
13. Mohan, K. Y. N., & Duraiswamy, P., (2016). On the optimization of GaN Schottky diode rectifier circuit topologies for high efficiency wireless power transmission. In: *2016 Asia-Pacific Microwave Conference (APMC)* (pp. 1–4). doi: 10.1109/APMC.2016.7931312.

CHAPTER 2

BUILDING SMART BOT DESIGNS FOR LIBRARY ASSISTANCE

S. VALAI GANESH[1], V. SURESH[2], S. GODWIN BARNABAS[1], and S. RAJAKARUNAKARAN[1]

[1]*Department of Mechanical Engineering, Ramco Institute of Technology, Rajapalayam, Virudhunagar, Tamil Nadu, India*

[2]*Department of ECE, National Engineering College, Kovilpatti, Thoothukudi, Tamil Nadu, India*

ABSTRACT

The primary goal of this endeavor is to locate the book and determine its availability. Using the mechanical rack that is available, the constructed design is tested at the college library. Based on the experimental setup, it can be noticed that the robot design complies with robotics laws and is environmentally sustainable for all library users. The next task will involve employing an autonomous robot to retrieve books from the rack and deliver them to the user. When the user inputs the required book, the robot locates the location and verifies its availability. It moves to the spot and uses the robotic arm to pick up the book. The robot uses its telescopic cylinder to pick up the book in the event that it is accessible on a separate rack shelf. The book is identified by its RFID tag and reader.

2.1 INTRODUCTION

Developing a Smart intelligence system to navigate the library users to the required book location with smart inbuilt techniques is the major work

Robotics and Automation in Healthcare: Advanced Applications. R. Thanigaivelan, Sanjay Singh, & Clement Christy Deepak C. (Eds.)

behind in this work. Physically Challenged people have the difficulty to reach the book location without any human assistance. The proposed idea has the Android application (equipped with Watson Assistant Chatbot) built in the mobile robot for finding the book location and the bot will navigate the user to the required location. If there are more than 100 of racks and 10,000 plus books are available in the library means, this smart bot will assist the user to guide them in a proper direction such that searching time will be reduced. Library management systems are available in the Bot for entry and return of books.

2.1.1 LIBRARY MANAGEMENT

Library is the place for reading or studying the collection of curated books available there. Daily, there are numerous users are entering and exit the library. For collecting the user's details, library management system is used in all libraries. And it is also used for entering the book which is take off from the library.

2.1.2 CHATBOT

Chatbot is capable of conversing with users through chat interface. Chatbot is a software application which is used for conducting online chat via text or voice-based system. It can work 24/7, unlike humans. Watson Assistant is an IBM cloud-based application used for making user friendly chatbot application. Watson Assistant has three features or steps to build the chatbot such as intents, entities, and dialogs. Intents are things which are used for finding the goal of the user or finding what the user wants? It starts with the symbol '#.' Entities allow us to capture specific values within the user's discourse. It starts with the symbol '@.' Dialogs determine the response based on the detected intents and entities. Dialogs are the main component of dialog skill which give response to user's queries by proper training given by the Watson model.

2.1.3 ARTIFICIAL INTELLIGENCE

The term artificial intelligence (AI) refers to any human-like intelligence displayed by a computer, robot or other machine. Normally AI refers to the ability of a computer or machine to think on its own and make decisions based on the problem that they have. AI has numerous applications like Drone system, Driverless Cars, etc.

2.1.4 MACHINE LEARNING (ML)

ML is a subfield of AI, getting computers to work without being directly programmed. ML covers the statistical part of AI. It has two types of learning approaches: Supervised and Unsupervised Learning.

2.1.4.1 SUPERVISED LEARNING

Supervised learning means observation and direct execution of a task or project. Teaching the model with the proper labeled data set. Example for supervised learning: classification and regression.

2.1.4.2 UNSUPERVISED LEARNING

Unsupervised learning means not supervising the data. Trains on the dataset and draws the conclusion on unlabeled data. Examples of unsupervised learning: clustering and density estimation.

2.1.4.3 DIFFERENT APPROACHES OF ML

The following are some examples of ML approaches:

i. **Regression:** Predicting continuous values.
ii. **Classification:** Predicting the class or category.
iii. **Clustering:** Group of similar cases.
iv. **Recommendation:** Based on the people's task "recommended" on the website related to the search.

2.1.4.4 LIBRARIES USED IN ML

The following are some of the libraries needed to do ML operations:

i. **Scipy:** It includes numerical algorithms, domain-specific toolboxes, and it is good for scientific computations.
ii. **Numpy:** Math library to work with 'n' dimensional arrays in Python.
iii. **MatplotLib:** Plotting 2D as well as 3D.
iv. **Pandas:** High level Python library and easy to data structures. It has many functions like importing, Manipulation, and analysis.
v. **Scikit Learn:** Free ML library for Python. It is similar to NumPy and SciPy.

2.1.5 DEEP LEARNING

Deep learning is a subfield of ML deal with algorithms caused by structure and union of the brain called artificial neural networks (ANNs). Convolutional neural networks and recurrent neural networks are examples of deep learning approaches.

2.1.5.1 TYPES OF DEEP LEARNING MODELS

There are two types of deep learning models are as follows:

i. Convolutional neural networks; and
ii. Recurrent neural networks.

2.1.5.2 CONVOLUTIONAL NEURAL NETWORKS

Convolutional neural networks (CNN) are mainly used for image processing, image detection and classification. CNN is used for processing the 2D matrix images. CNN comprises of neurons with learnable weighs and bias. Each neurons receives numerous inputs, and all take a sum over and then the results are processed.

2.1.5.3 RECURRENT NEURAL NETWORKS

Recurrent neural networks (RNN) are a class of neural networks which helps during modeling the sequenced data. RNN functions similar to the human brain. Prediction outcomes generated by recurrent neural networks, which are not achievable with other algorithms.

2.2 RELEVANT WORKS

Microsoft access is used in the database, for finding the user details and books available in the library [1]. Robotic arm is used to pick and place the book with the aid of a barcode reader, IR Sensor. Processing software is used for sending the commands to the Arduino controller through GUI. If the user finds the availability of a book, then the signal is sent to the robotic arm through a wireless network. Library administration system is being assessed by the human individual for bill payments [2], book tracking, etc., Usually in all libraries, need librarian to pick the book if the floor size is large. To avoid these troubles, automation is done in the library for delivering the books with the aid of robotic arm. This work focuses the pick and place of book using mechatronics system.

An attempt to design the robotic library system with the aid of AI. Functioning of robot based on ARM microprocessor, motor drive circuit with 5 DOF, Wi-Fi, and GPS based communication protocol [3]. The verification of library books is controlled by RFID. In this library issuance system, the previous capable reader's verification review reports have been considered for recommending suitable books to the deserving new users and the issuance of books or periodicals is based on the user's decision.

Robust lane detection methods are also used. Extraction of lane edge in traffic scenario is found by adapting thresholding strategy to binarize the gradient image and trace edges [4] by using local gradient information. Further, through inverse perspective mapping the endpoints of extracted lines to the world coordination, we can combine the extracted lines from different cameras. Finally, the lane could be detected by matching two points instead of two parallel lines in parameter space. An autonomous library assistant robot [5] that was developed and implemented in the real time function. The robot development and its functioning are based on image processing technique combined with Odometry and sonar data which is checked by the Kalma filter.

The robotic assistant which mainly helps the library user [6]. According to the user's input, the robot will search the book. If found, it will return to the user as early as possible. The robotic system integrates object recognition, visual grasping, and other advanced capabilities.

Based on the review, it is identified that library automation is done partially not fulfilled. So, the proposed work focused that the Watson assistant chatbot is used for finding the book availability and it is interfaced with the line follower robot which help the user to navigate to the required book location.

2.3 COMPONENTS USED

2.3.1 BODY, LEG, AND HEAD ASSEMBLY

The part of robot can't create any degree of freedom. It only supports the user to navigate to the required book location. Under the robot, the motion control kit is available to control the movement of robot (Figure 2.1).

FIGURE 2.1 Assembled robot model–model was created and assembled using Autodesk fusion 360 [7].

2.3.2 SENSORS AND ACTUATORS

Ultrasonic Sensor is an electronic device used for obstacle detection. It measures the distance of the object by emitting the ultrasonic sound waves and it converts the sound waves into the electrical signal. DC Motor is one of the types of rotary motor that converts the electrical energy into the mechanical energy. All of the motor type has the same mechanism as electromechanical and electronic to change the direction of current. It is based on the principle of electromagnetic induction. Servomotors are electronic equipment, and it is used for angular displacement. It needs a separate controller that needs the program to control the servomotor. Servomotors are used in applications such as robotics, CNC application, and Automated applications.

2.3.3 *MYRIO HARDWARE*

myRIO is an embedded application board made by the company called NI Instruments. myRIO has the I/O in the form of MXP AND MSP connectors. It is used to develop applications that uses its onboard Field Programmable Gate Array and microprocessor. It requires LabVIEW software to make the application.

2.3.4 *OTHER COMPONENTS*

2.3.4.1 L293D MOTOR DRIVER

L293D is commonly used motor driver with 16 pin configurations. A single L293D IC has the capability to control the motion of two DC motor at a time. The L293D IC work on the principle of Hall H-Bridge. It can also be used for controlling the servomotors.

2.3.4.2 NUT AND BOLT

Nut and bolt are used for the joining process (temporary joining process). The bolt comprises of a head and a cylindrical body with screw threads along a portion of its length. The nut is the pairing member of the bolt, having internal threads to match those of the bolt. Washers are often used to prevent loosening and crushing.

2.3.4.3 LINE FOLLOWER CHASSIS

Chassis is a base frame of the line follower robot. Chassis are used to hold or place the other electrical components which are used to make the line follower robot.

2.3.4.4 3D PRINTER

3D printing is an additive manufacturing technique, and it constructs the three-dimensional object from the CAD model of the object that we imported in the machine. The printing technique used in the 3D printing is a fused deposition method. Design the 3D model and convert the file format to .stl file and import that file in the 3D printer and finally print the model using 3D printing technique. Material used for making the 3D models are polylactic Acid (PLA), Acrylonitrile Butadiene Styrene (ABS), Polyethylene Terephthalate Glycol (PETG), etc.

2.3.4.5 JUMPER WIRE

Jumper wires are generally used with breadboards and other prototyping tools in order to make it easy to change a circuit as needed. In this project, jumper wires are used to make the electrical circuit.

2.4 CHATBOT DESIGN

2.4.1 INTRODUCTION

Chatbot is a tool used for conducting online chat and it is used for clarifying the user's query at any time (24/7). Watson Assistant is an IBM's product that will help you to build, train, and deploy the interacted chatbot in any websites or any custom-based applications and it is used for making the chatbot. The salient features available in the Watson assistant are Intents, Entities, and Dialogs of the dialog skills.

2.4.2 INTENT

Intents are the things which are used for finding the goal of the user or finding what the user wants? It starts with the symbol '#.' Examples for Intents: greetings, thankyou.

2.4.3 ENTITIES

Entities allow us to capture specific values within the user's discourse. It starts with the symbol '@.'

Examples for entities: @book, @department.

2.4.4 DIALOGS

Dialogs determine the response based on the detected intents and entities. Dialogs are the main component of dialog skill which give response to user's queries by proper training given to the Watson model.

Example conversion: greetings: Hi
Response: Hi there, How can I help you?

2.4.5 STEPS TO CREATE THE CHATBOT

- **Step I:** Creating the account in the IBM Cloud.
- **Step II:** Create the instances in the Watson Assistant.
- **Step III:** In services, create the Watson assistant service. Create instance for free version. Free version only has the access to create the chatbot which has the limit of only 10,000 messages. If it goes above the limit, it asks Lite plan.
- **Step IV:** Create Intents by clicking the button available at the top corner. It has the feature that export and import of intents are done by adding or downloading the excel file in.csv format.
- **Step V:** Create entities by clicking the button available at the top corner. It has the feature that export and import of entities are done by adding or downloading the excel file in.csv format.
- **Step VI:** Adding dialog to the dialog skills. The dialog for the identified intents and entities are given here in the flow chart format. If the condition satisfies, it gives the output. Otherwise, it simply says "I can't understand, Please rephrase the sentence."
- **Step VII:** Deployment of Chatbot in the required location like application, website, etc. Watson Assistant tool has the numerous

features like integrate the chatbot with webhooks, Autocorrection, Disambiguation, Irrelevance detection and intent detection.

2.4.6 *WORKFLOW OF CHATBOT* (Figure 2.2)

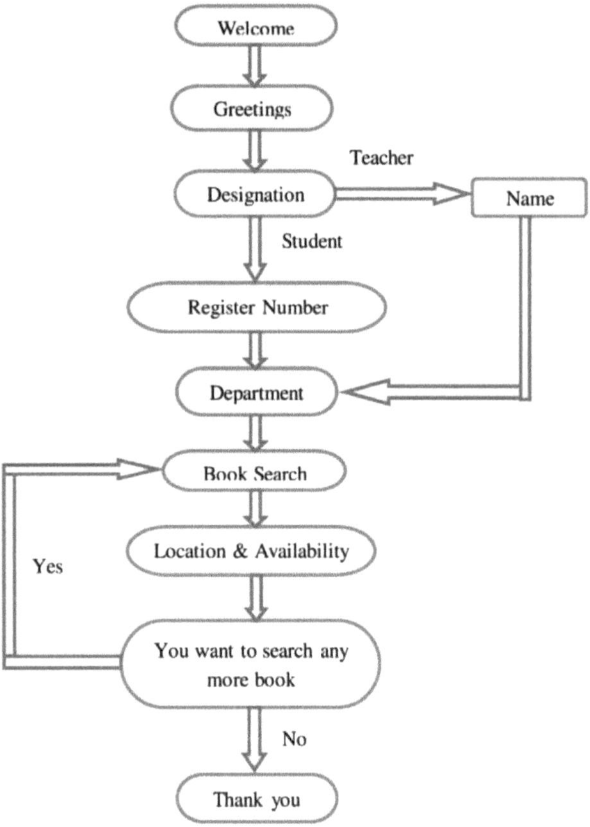

FIGURE 2.2 Workflow of chatbot–developed using IBM Watson assistant.

2.4.7 *SAMPLE OUTPUT*

In our work, we proposed the chatbot for finding the book availability in the general library of college. Mechanical rack of the library is taken and tested for making the curated results. The taken books are listed in Table 2.1 for reference. Initially, chatbot greetings the user and it asks the designation of the user. According to the user's reply, its response. If the library has the searched book, then it replies to the book location. If not there, it replies "The book that you need is currently not available in the library." Finally, it thanks the user, and it replies to the welcome message. The image shows the sample output of the chatbot what we have generated (Figures 2.3(a)–(c)).

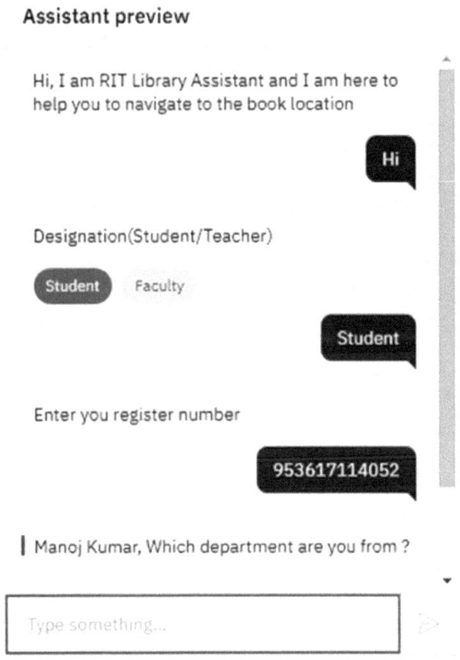

FIGURE 2.3(a) Sample output from Chatbot 1 [8].

2.5 METHODOLOGY

2.5.1 RELATED WORKS

Collecting relevant:

- Literature related to automation in library management system.
- Performance of robot using the mechatronics instruments for book pick and place purpose.
- Watson assistant ChatBot.

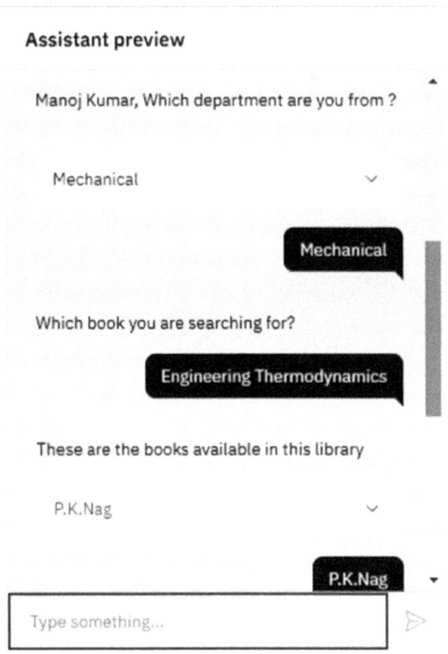

FIGURE 2.3(b) Sample output from Chatbot 2 [8].

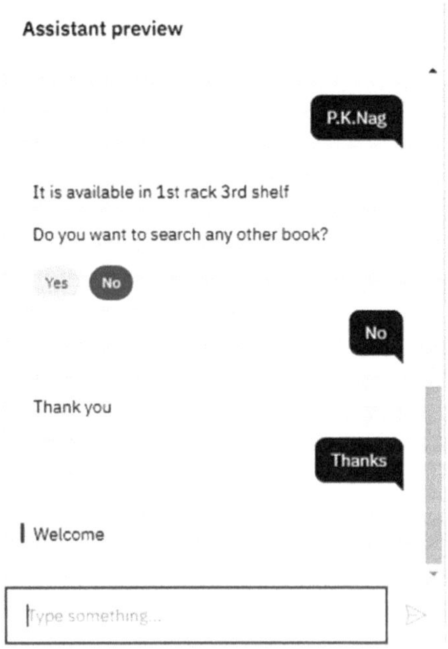

FIGURE 2.3(c) Sample output from Chatbot 3 [8].

TABLE 2.1 Book List [10]

SL. No.	Book List	Shelf	Count
1.	Engineering Thermodynamics	3	60
2.	Thermodynamics	3	22
3.	Strength of Materials	1	48
4.	Aircraft Maintenance and Repair	6	3
5.	Automobile Engineering	6	26
6.	Reliability Engineering	7	21
7.	Product Design and Development	7	18
8.	CO_2 Emission Mitigation	7	5
9.	Introduction to Biofuels	7	4
10.	Bio-Process Engineering Principles	7	3
11.	Engineering Heat and Mass Transfer	4	3
12.	Fundamentals of Heat Exchanger Design	4	2
13.	Heat and Mass Transfer	4	173
14.	Heat Transfer	4	33
15.	A Course in Heat and Mass Transfer	4	3
16.	Engineering Heat and Mass Transfer	6	3
17.	Jigs and Fixtures	6	11
18.	Pneumatics-Concepts, Design, and Applications	6	6
19.	Fundamentals of Compressible Flow	6	12
20.	Modern Compressible Flow	6	6
21.	Aircraft Performance Theory and Practice	6	1
22.	Aircraft Electricity and Electronics	6	3
23.	Machines and Mechanisms	5	5
24.	Theory of Machines and Mechanisms	5	52
25.	Elements of Workshop Technology	9	35
26.	Workshop Technology	9	8
27.	Steam Tables with Mollier Diagram	10	105
28.	Heat and Mass Transfer Data Book	11	150
29.	Design Data Book	12	157
30.	Mechanical Metallurgy	8	5
31.	A Textbook of Production Technology	8	9
32.	Product Design and Manufacturing	8	10
33.	Manufacturing Planning and Control	8	3
34.	Elements of Manufacturing Process	8	15

2.5.2 MODELING

Robot body has many parts which need to be done with more accurate manner. It contains body, leg assembly and head assembly. The modeling of the Robot is done in the software named Autodesk Fusion 360.

2.5.3 WATSON ASSISTANT CHATBOT

The chatbot is helpful for clearing the user's query at any time (24/7). Watson Assistant is a tool which is used for making the chatbot with the aid of the features like intents, entities, and dialogs which are available in the IBM Cloud.

2.5.4 DEPLOYMENT OF CHATBOT IN APPLICATION

The developed chatbot is deployed in the mobile application. The mobile application is created using the Android Studio [11].

2.5.5 FABRICATION

Based on the proposed design, the fabrication is done in two specific stages:

- **Stage 1:** The proposed and assembled design is sliced into various parts and fabricated using 3D printing by setting up the optimal specifications.
- **Stage 2:** The fabricated components are assembled using glue which is specially available for 3D printed materials.

2.5.6 INTEGRATION OF ROBOT FOR MOTION CONTROL

The electronic circuits are integrated with the robot for motion control. Library management system (check-in and check-out purpose) are designed using the LabVIEW Software by using the features like constant, control, and indicator. The LabVIEW Software is interconnected with the myRIO hardware for motion control of library assist robot.

2.5.7 DESIGN OF LIBRARY MANAGEMENT

The Developed library management system has the options like resource-in and resource-out of the library book, Mail Intimation (Reminder) for the returning the book. If the user wants to take the book, then click on the resource-in button. After that, it asks the user details like register number and date of birth. Then select the book list from the available resource. If the user wants the book location, then click the location button. If the user clicks the location button, it replies as "it is available in the 2nd rack 3rd shelf." If the user wants to return the book, click resource-out button. Then it asks to select the book. Then we return the book without any guidance from the librarian. Developed library management system created in LabVIEW software (Graphical Programming) for resource-in purpose.

2.5.7.1 LABVIEW INTRODUCTION

Laboratory Virtual Instruments Engineering Workbench is created by the National Instruments. It has many special features and most noted feature is the graphical programming environment. It has wider range of applications like signal processing, image processing, machine vision. The major steps in the creation of graphical programming include:

- Operate the instrumentation program;
- Control selected hardware;
- Analyze acquired data;
- Display results.

2.5.8 WORKING OF LIBRARY MANAGEMENT SYSTEM

Figure 2.4 shows the user interface of developed library management system which created by using the constant, control, and indicator.

If the user wants to take the book from the library, select the resource-in button available in the user interface. After selecting, the figured page (Figure 2.5) will appear on the screen.

If the user clicks the "OK" button, the catalog of the books is listed in the table for user reference. They may select different author books for the same subject (Figure 2.6).

FIGURE 2.4 User interface [9].

FIGURE 2.5 Book selection pop-up.

FIGURE 2.6 Catalog of books.

After selecting the required book from the available list of books, it shows the pop-up as "Do you want the book location?" If the user clicks the "Yes" button, the integrated and embedded robot will navigate the user to the required book location (Figure 2.7).

FIGURE 2.7 Book location pop-up.

After the navigation to the required book location, it asks the user for confirmation pop-up as "Do you want to take this book?" If the user clicks "Yes," then it finishes the entry in the library management system (Figure 2.8).

FIGURE 2.8 Confirmation pop-up.

If the user click "Yes" button, it asks the register number and date of birth for check-out purpose. After the entry of user details, it enters the details in the library management system (Figure 2.9).

FIGURE 2.9 User details.

After entering the user's details like register number and date of birth, it shows the return date of the book that the user wants (Figure 2.10).

FIGURE 2.10 Return pop-up.

The developed library management system created in LabVIEW software (Graphical Programming) for resource-in purpose is shown in Figure 2.11.

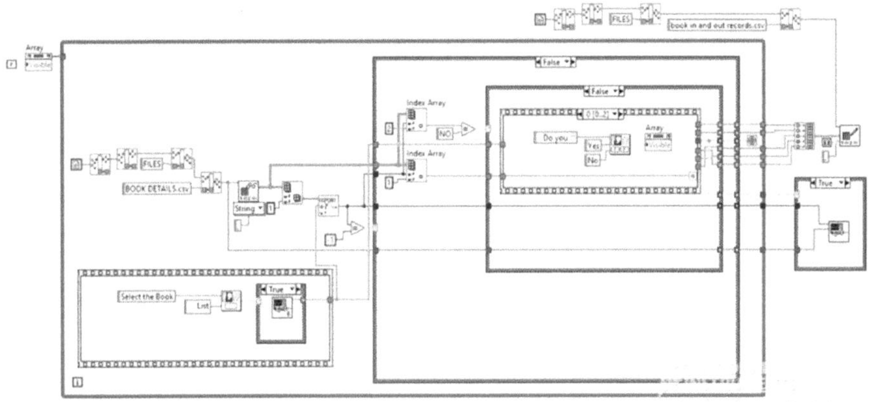

FIGURE 2.11 Resource-in graphical programming [9].

The developed library management system created in LabVIEW software (Graphical Programming) for resource-out purpose is shown in Figure 2.12. The user may return the book without any human assistance. It asks the book which the user want to return.

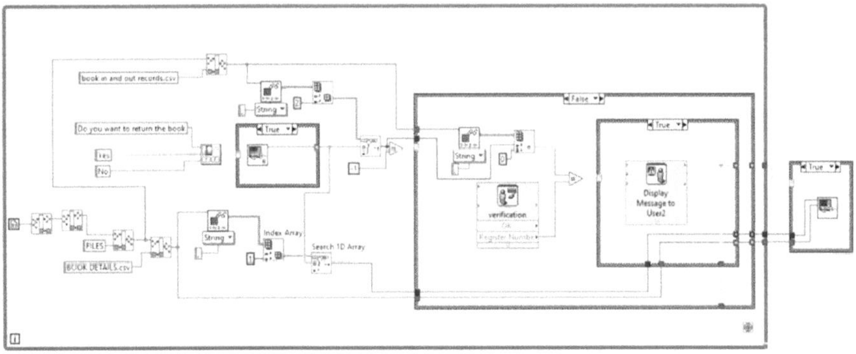

FIGURE 2.12 Resource-out graphical programming.

Resource-out purpose graphical programming will alert the user to return the book. It will send the e-mail remainder before two days of the return date (Figure 2.13).

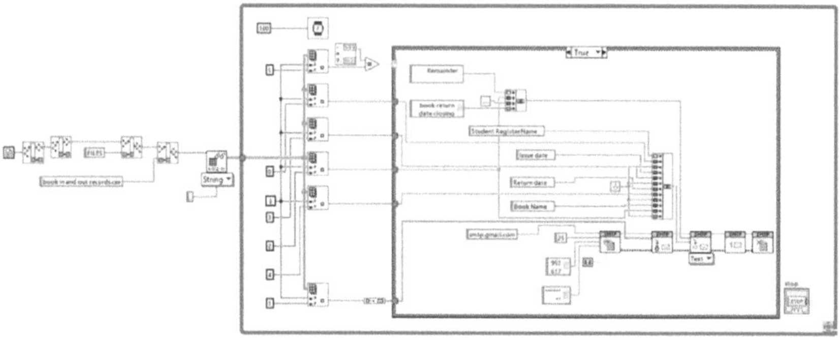

FIGURE 2.13 Remainder mail graphical programming.

2.5.9 *FABRICATED MODEL*

The modeled assembled robot is fabricated in the 3D printer available in the college laboratory. The modeled robot is sliced into three components head, leg, and body, and then fabricated in the 3D printer.

2.5.9.1 OPERATING PARAMETERS OF 3D PRINTER

The operating parameters of 3D printer are as in Table 2.2.

TABLE 2.2 Operating Parameters

SL. No.	Operating Parameters	Value
1.	Layer	PLA
2.	Layer height	0.25
3.	Initial layer height	0.15
4.	Line width	0.35
5.	Wall thickness	0.8
6.	Top/bottom thickness	0.8
7.	Support density	20
8.	Pattern	Zigzag
9.	Print speed	50

The navigation of user to the required book location will be obtained by taking the input that the user enters in the library management system. The robot is programmed based on the distance of the book location from the start position.

2.5.10 EXPERIMENTAL VERIFICATION AND REPORT SUBMISSION

The fabricated robot is experimented in the general library of the college. Chatbot is tested by searching the different books and by different users. Mechanical rack available in the library is taken and the robot is tested for finding the book availability in the particular rack (Figure 2.14).

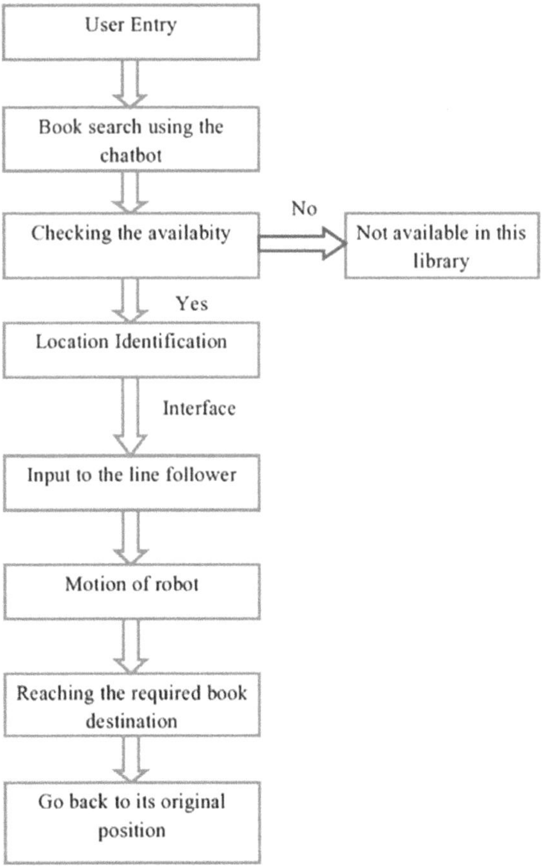

FIGURE 2.14 Flow of proposed model.

2.6 ADVANTAGES AND APPLICATIONS

The developed system has the following advantages:

- The library user can use the library without any human assistance.
- It reduces the human assistant by introducing the robot with minimal cost.
- The user can easily find the location and availability of the book. It reduces the time wasted in searching for the book physically.
- It reminds the user before the due date.

The developed system has the following applications:

- Used in library for navigating to the required book location. Visually Impaired people also get benefitted.
- Fabricated robot is also used for sorting the waste in dump yard.
- It will also be used for supermarkets, malls, shops, etc., by reducing the human assistance.

2.7 RESULTS AND OUTCOMES

The fabricated robot, along with the integrating circuits for the line following robot, is tested in the general library of the college. The library has different racks for different departments. The mechanical department rack is taken, and the books in the racks are included in the chatbot. The prepared chatbot is tested by sharing the preview link to the 20 different users. The chatbot integrated with the robot is tested, and results are taken successfully. The outcomes from this work are the development of a smart intelligence system for making library automation, easing the process of library workers, and automating the library management system to find book locations and availability.

2.8 CONCLUSION AND DISCUSSION

This work is mainly focused on finding the book availability and its location. The fabricated design is tested in the college library by taking the mechanical rack available in the library. From the experimental setup, it is observed that the robot design is not violating the laws of robotics and it is eco-friendly to all users who use the library. The future work is to pick the book from the rack and place it to the user by using the autonomous robot. If the user gives the needed book as the input, then the robot checks the availability and finds the location. It moves to the location and picks the book using the robotic arm. The telescopic cylinder is used in the robot to pick the book if the book is available on a different shelf of the rack. The RFID tag and reader are used for identifying the book.

KEYWORDS

- **artificial intelligence**
- **artificial neural networks**
- **convolutional neural networks**
- **machine learning**
- **polyethylene terephthalate glycol**
- **recurrent neural networks**

REFERENCES

1. Pujari, T. S., & Deosarkar, S. B., (2017). Design of intelligent and robotic library system. In: *2nd IEEE International Conference on Recent Trends in Electronics, Information Communication Technology (RTEICT)* (pp. 1903–1908).
2. Gade, A., & Angal, Y., (2016). Automation in library management using LabVIEW. *In Proceedings International Conference on Computing Communication Control and Automation (ICCUBEA)* (pp. 1–15).
3. Sreejith, M., Joy, Abhishesh, P., Ryuh, B., & Kumar, V. S., (2015). Conceptual design of a Wi-Fi and GPS based robotic library using an intelligent system world academy of science, engineering, and technology. *International Journal of Computer, Electrical, Automation, Control, and Information Engineering, 9*, 2511–2515.
4. Sun, T., Tang, S., Wang, J., & Zhang, W., (2013). A robust lane detection method for autonomous car-like robot. *Fourth International Conference on Intelligent Control and Information Processing (ICICIP)* (pp. 373–378).
5. Behan, J., & O'Keeffe, D. T., (2008). The development of an autonomous service robot. Implementation: "Lucas"—The library assistant robot. *Intelligent Service Robotics, 1*, 73–89.
6. Ramos-Garijo, R., Prats, M., Sanz, P. J., & Del, P. A. P., (2003). An autonomous assistant robot for book manipulation in a library. *SMC'03 Conference Proceedings. 2003 IEEE International Conference on Systems, Man, and Cybernetics* (Vol. 4, pp. 3912–3917). Conference Theme – System Security and Assurance (Cat. No. 03CH37483).

7. *Autodesk Fusion 360Software*. https://www.autodesk.com/products/fusion-360/overview (accessed on 20 November 2023).
8. *IBM Watson Assistant*. https://www.ibm.com/in-en/products/watson-assistant (accessed on 20 November 2023).
9. *NI LabVIEW Software*. https://www.ni.com/en-in/shop/labview.html (accessed on 20 November 2023).
10. *Ramco Institute of Technology*. https://www.ritrjpm.ac.in (accessed on 20 November 2023).
11. *Android Studio App Development*. https://developer.android.com/studio (accessed on 20 November 2023).

CHAPTER 3

EVALUATING THE BIOMECHANICAL EFFECTS OF TITANIUM DENTAL IMPLANTS OF DIFFERENT SIZES ON STRAIN DISTRIBUTION USING THE STRAIN GAUGE METHOD

DURAISAMY VELMURUGAN[1], THANGAVEL YUVARAJ[1], and RAJENDRAN SELVAPRIYA[2]

[1]*Department of Mechanical Engineering, Muthayammal Engineering College, Rasipuram, Namakkal, Tamil Nadu, India*

[2]*Department of Civil Engineering, Muthayammal Engineering College, Rasipuram, Namakkal, Tamil Nadu, India*

ABSTRACT

The experimental approach utilized to investigate the biomechanical properties of two implants with various diameters is described in this article. The 4.2 mm and 5 mm implant diameters were described, and the recorded readings were taken note of. The stress/strain dispersion around the bony-implant region is significantly influenced by the diameter of the dental implant. In this study, the mesial and distal sides of the bone block were examined.

3.1 INTRODUCTION

Dentistry has gained recent attention worldwide as it concentrates on people's well-being by boosting their morale in terms of how they look

Robotics and Automation in Healthcare: Advanced Applications. R. Thanigaivelan, Sanjay Singh, & Clement Christy Deepak C. (Eds.)

and feel about themselves. The substitution for the loss of natural teeth is necessary to enhance the appearance, self-esteem, and happiness and this is achieved through a dental implant, which is considered as the finest and ideal option in modern days. The success rate of dental prosthesis is found to be high worldwide according to the available data, and this can be attained through proper design and correct placement of the implant into the jaw. However, the failure of the implants is still reported; this is happening due to lack of knowledge in stress/strain distribution characteristics connected to the mandible/maxilla and implant biomechanical behavior [1]. Mechanical intricacies and failures have been stated in the interim of prosthodontic treatment [2]. The biomechanical study of a dental implant began in the late 1980s, and using various bioengineering tools, efforts have been made to study the impact of different factors on biological issues. The failure of the implantation is influenced by a number of factors; among them, the biomechanics of the implant is a crucial factor of concern [3]. Different kinds and magnitudes of forces are acting on the implant in various directions, and these loads are transmitted to the adjoining tissues. The manner in which the load is transferred to the adjoining biological tissue is very important, since it helps to maintain the implant-bone interface, and also the load transferring mechanism helps to maintain the implant longevity. Additionally, the load acting on the implants produces undesired stresses/Strains in the upper and lower jaw bones, and this may lead to implant failure [4]. At this junction, understanding the concept of biomechanics has gained much attention in analyzing the influence of loads acting on the implant and the stresses/strains induced in the implant and the surrounding bone. The latest advancement in dental Implantology has a great impact on Osseo integrated dental implants. However, more research in this area is required to meet the challenges that arise in dental implantation procedures. The problem may occur in clinical applications since the implants and the natural teeth act with different biomechanical aspects. The way the load and stress dispersion to the bone and implant surroundings is important, it is critical to know the implant survival life. Hence the field of stress analysis in dentistry, extensively at implant, implant-bone interface, gained popularity to predict the stress/strain generated around this region, so that the area of maximum concentration of stress can be observed in such a way that the implant survival life also comes under the surveillance. In dentistry, several methods are utilized to examine stress/strain in the region of bone-implant interface. These methods are categorized as experimental, analytical, and computational methods by the way of photoelasticity, Strain Gauge and Finite Element Analysis (FEA) approaches.

As detailed in the literature section, geometry of the implants plays a crucial role in determining the biomechanical behavior at bony segment. The designing of dental implant has long been considered as a critical parameter for dental implant treatment and also its success. There are a greater number of reported studies which indicates that stress/strain in the bony region of implant depends on the type of the implant dimensions preferred. Stress/strain attention on the implant-bony region is firmly influenced by the implant diameter than the implant length considered. Diameter of the implant leads to produce more stress in the cortical bone than the cancellous bone, and this geometry of the implant also affects the level of interfacial stress conditions. The usage of shorter implants (diameter of the < 3.7 mm) in dentistry is when the alveolar crest ridge is thin, restoring a tooth with a minimum-dimensions and also where limited space exists in areas such a sinter-radicular region. The interaction occurring between the bone-implant regions is also important since this interaction level influences the strain value around the implants. The strain in the implant surface area is a much essential factor to examine since it influences the bone stress too. The influence of implant diameter on peri-implant region has been examined by various researchers using the numerical analysis approach with some limitations. However, in those studies, bonded contact is assumed between the bone-implant interfaces which shows the perfect osseointegration state, but this condition is not possible to be achieved in real clinical setting. In order to overcome these drawbacks, it is better to concentrate on the qualitative approach rather than the quantitative approach.

Strain Gauge method is utilized to evaluate stresses *in vivo* and *in vitro* in dental prostheses, implants, and natural teeth. From the view of biomechanical consideration, the bonding between the implant and bone should minimize the stress/strain level which is generated around the bone implant interface, so that the fatigue and micro-strain induced at the surfaces can be avoided. The biological and technical problems are still in line, which affect the life of the implants. Hence, these factors should be addressed properly. Therefore, the aim of the present chapter is to study the biomechanical behavior of the two implants with different diameters, in vitro model, by using the strain gauge analysis method.

3.2 EXPERIMENTAL METHODS AND MATERIALS

3.2.1 STRAIN GAUGE

A strain gauge is a sensor-based device whose electrical resistance varies accordance to the applied force. It changes force, weight, and pressure into a change in electrical resistance, which can then be measured. It is considered as one of the significant sensors used to measure the mechanical quantities. As the name shows, it is used to measure the strain, which is the technical call containing both tensile and compressive strain. Hence it is used to measure both extensions well as contraction in strain value. Based on the applications, strain measuring techniques are classified as mechanical, optical, and electrical strain gauges. There are some strain gauges called extensometer, used to measure strain value in both extension and shorten which occurs mechanically. In the case of optical strain gauges, the strain on the material surfaces induced by the applied force is determined using the technique called stereo photogrammetry. Electrical strain gauges work based on the change in electrical resistance when the external load is acting on the object.

3.2.2 STRAIN GAUGE IN IMPLANT DENTISTRY

It is a small electrical sensor used to measure deformation where they are employed. The stored electrical signals are sent to the DAQ, and converted into digital signals, read by the computer. The gauges are capable of measuring deformation value accurately when the object is under stress. The use of the strain gauge to examine the stress present in the implant supports the clinical certainty. However, no complete information exists about the model which is to be used to perform the biomechanical study using strain gauges. According to the source available, few of them employed strain gauges directly on prosthetic parts [5] and few others used SG on the material which shows similar bone properties [6].

3.2.3 SELECTION OF MATERIALS

Biomechanical experimentation and computer simulation are the available research tools in dental implant biomechanics for the past few years. Human cadaver bone has been used for experimental and verification purpose due to its close properties' *in vivo* characteristics of the bone. However, handling the cadaver bones poses lot of difficulties, hence finding an alternative material such as synthetic bone has gained popularity in the recent times. Several synthetic bone materials are available, though polyurethane foam has been utilized in dental implant biomechanics, especially for stress/strain measurements around the implant surface and also bone-implant interface. The foams are produced through the polymerization process, and the outcome of the product is closed cell structure which is different from cancellous bone in terms of open porosity. However, the reliability and uniformity in the material properties aid to fabricate rigid polyurethane structure, which is the ultimate model for dental implant testing. Therefore, in this work, solid rigid polyurethane foam was utilized in custom laminated bone blocks for experimental purpose. The typical solid rigid polyurethane foam used in biomechanical experimentation is shown in Figure 3.1.

FIGURE 3.1 Custom laminated bone block.

3.2.3 PROPERTIES OF SOLID RIGID POLYURETHANE FOAM

It executes tremendous mechanical property, light weight in structure, but strong resistance against the applied load. Also, it has a density value from 30 kg/m^3 to 100 kg/m^3 based on the applications involved. It can increase the density value up to 700 kg/m^3 when the extreme level of mechanical load is considered. The mechanical property of the polyurethane foam is also found to be good at low densities. Based on the densities, the stress value lies from 40 to 900 kPa, and the range of shear stress values lies from 120 kPa to 450 kPa. Solid rigid polyurethane foam gives only ±10% deviation in density ranges and most probably that offers the uniformity in the structure. It is used as an alternative material choice for human cancellous bone. Moreover, it does not imitate the properties of the cancellous bone, but it can provide some stable properties with close resemblance to the human cancellous bone.

3.2.4 IMPLANT IN BONE BLOCK

The details of experimental setup used in this study are mentioned below. Dental implants with different diameters have been selected for this study. Implants are inserted into the polyurethane foam block (SAW BONES, USA) as shown in Figure 3.2. The custom laminated bone block made of solid rigid polyurethane foam and short fiber filled epoxy sheets are used in this study. This bone block is used to replicate the mandible bone properties. Solid rigid polyurethane foam represents the human cancellous bone, and epoxy sheet represents the cortical bone properties. The detailed dimensions and material properties of the bone block are listed in Table 3.1.

FIGURE 3.2 Implant in bone block.

TABLE 3.1 Material Properties of the Bone Block

Material	Dimension	Density (Kg/m^3)	Elastic Modulus (GPa)
Cortical bone	2 mm in thickness	–	16.7
Cancellous bone	30 × 41 × 42 mm in thick	640	0.759

3.2.5 BONDING OF THE SG

The bonding of the SG on the bone block is a very crucial task, and it requires much careful concentration for measuring the strain. The material surface finish is important for bonding the SG. Two rectangular rosette strain gauges with six grids are bonded at mesial and distal side of the bone, as shown in Figure 3.3.

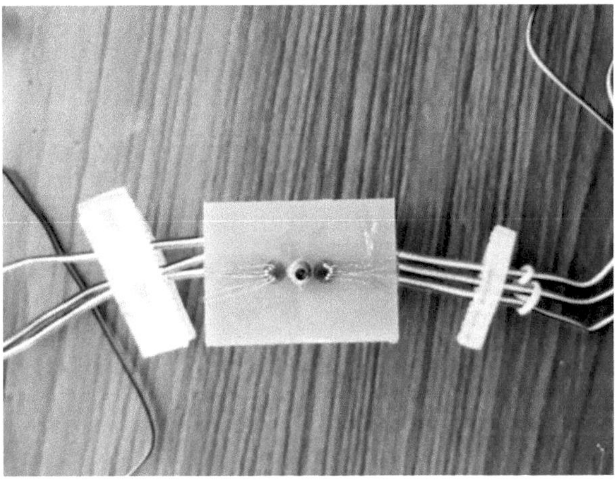

FIGURE 3.3 Bonding of strain gauge.

The strain gauges are bonded, and wirings are performed as shown in Figure 3.3 according to all the relevant regulations and guidelines. Cyano-acrylate cement was preferred for bonding the strain gauges on the cortical region of the bone model. The wiring on the SG is shown in Figure 3.4.

3.2.6 LOADING CONDITIONS AND MEASUREMENT

The utilization of the SG analysis in dentistry is considered only in low loading applications, which varies from 20 to 300 N. The universal testing machine (UTM) and custom-made load application devices are used to apply the load. But UTM is suitable for the high loading conditions, and using this in dentistry may cause some troubles while measuring the strains, since small values are often associated with the dental implantology filed. Hence, jig fixtures were used, as shown in Figure 3.5, to apply the load. After the loading protocol was over, the signals corresponding to that measured by the rectangular rosette strain gauge were sent to a DAQ, and the corresponding signals were analyzed using suitable software.

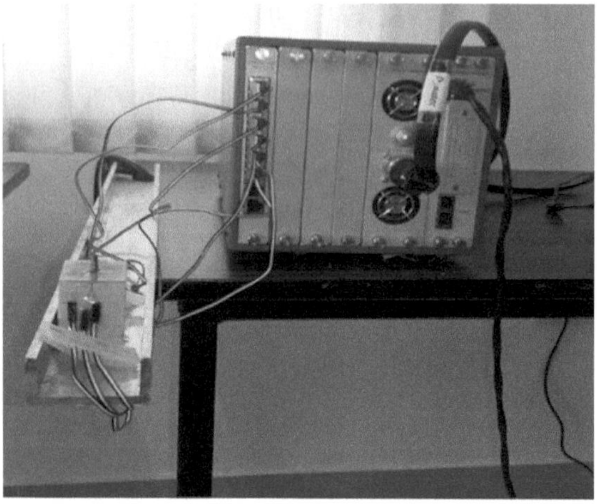

FIGURE 3.4 Wiring of the strain gauge.

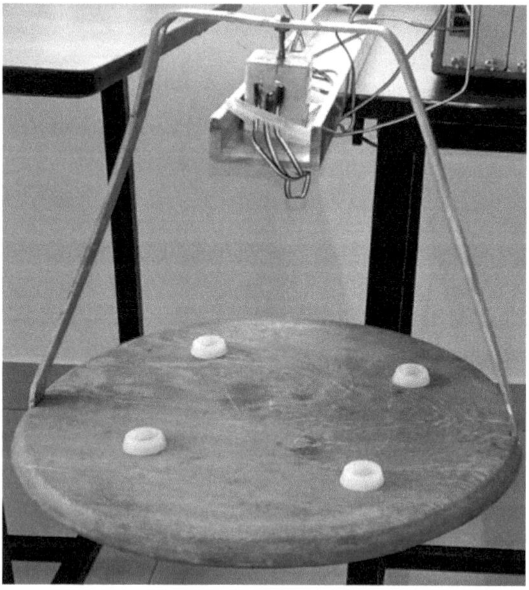

FIGURE 3.5 Jig device for applying the load.

There are four rectangular rosette strain gauges bonded in each side of the specimens for two different diameters of the implants. Strain at grids 1, 2, and 3 are corresponding to rosette 1, strain at grids 4, 5, and 6 are measured from rosette 2, and this procedure holds good for another bone specimen (implant with 5 mm diameter) also. The vertical load of 50 N, 100 N, and 150 N is applied to the implant and the complete experimental setup is shown in Figure 3.6.

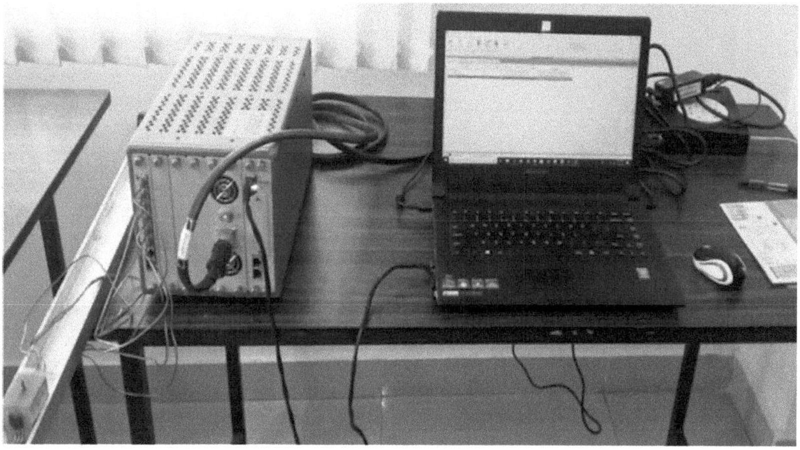

FIGURE 3.6 Complete experiment setup.

The following formulas related to rosette strain gauge are used to calculate maximum strain and the stress as well:

$$\epsilon_p = \frac{\epsilon_1 + \epsilon_3}{2} + \frac{1}{\sqrt{2}}\sqrt{(\epsilon_1 - \epsilon_2)^2 + (\epsilon_2 - \epsilon_3)^2}$$

$$\epsilon_q = \frac{\epsilon_1 + \epsilon_3}{2} - \frac{1}{\sqrt{2}}\sqrt{(\epsilon_1 - \epsilon_2)^2 + (\epsilon_2 - \epsilon_3)^2}$$

$$\sigma_p = \frac{E}{2}\left[\frac{\epsilon_1 + \epsilon_3}{1-\nu} + \frac{\sqrt{2}}{1+\nu}\sqrt{(\epsilon_1 - \epsilon_2)^2 + (\epsilon_2 - \epsilon_3)^2}\right]$$

$$\sigma_q = \frac{E}{2}\left[\frac{\epsilon_1 + \epsilon_3}{1-\nu} - \frac{\sqrt{2}}{1+\nu}\sqrt{(\epsilon_1 - \epsilon_2)^2 + (\epsilon_2 - \epsilon_3)^2}\right]$$

where; Strain from grid 1; strain from grid 2; strain from grid 3; maximum principal strain; minimum principal strain; maximum principal stress; minimum principal stress.

The experiment has been repeated several times to obtain the accuracy in the values. The strain values corresponding to the grid 1, 2, and 3 for rosette 1, and corresponding to 4, 5, and 6 for rosette 2 are listed in Tables 3.2 and 3.3 for both 4.2 mm and 5 mm diameter of the implant. The variations in signals also observed when the vertical load is increased 50 N, 100 N, and 150 N.

TABLE 3.2 Strain Values for 4.2 mm Diameter of the Implants at Grids

Assignment	**R1**	**R1**	**R1**	**R2**	**R2**	**R2**
Reduction method	**Strain (G1)**	**Strain (G2)**	**Strain (G3)**	**Strain (G4)**	**Strain (G5)**	**Strain (G6)**
Seconds elapsed	**[1, 2, 3] Strain (Grid 1)**	**[1, 2, 3] Strain (Grid 2)**	**[1, 2, 3] Strain (Grid 3)**	**[4, 5, 6] Strain (Grid 1)**	**[4, 5, 6] Strain (Grid 2)**	**[4, 5, 6] Strain (Grid 3)**
0.1	3	2	2	2	2	2
0.2	3	2	2	2	2	2
0.3	3	2	2	2	2	2
0.4	3	2	2	2	2	2
0.5	3	2	2	2	2	2
0.6	3	2	2	2	2	2
0.7	3	2	2	2	2	2
0.8	3	2	2	2	2	2
0.9	3	2	2	2	2	2
1	3	2	2	2	2	2
1.1	3	2	2	2	2	2
1.2	3	2	2	2	2	2
1.3	3	2	2	2	2	2
1.4	3	2	2	3	2	2
1.5	3	2	2	3	2	2
1.6	3	2	2	2	2	2
1.7	3	2	2	3	2	2
1.8	3	2	2	3	2	2
1.9	3	2	2	2	2	2

TABLE 3.2 *(Continued)*

Assignment	R1	R1	R1	R2	R2	R2
Reduction method	**Strain (G1)**	**Strain (G2)**	**Strain (G3)**	**Strain (G4)**	**Strain (G5)**	**Strain (G6)**
Seconds elapsed	**[1, 2, 3] Strain (Grid 1)**	**[1, 2, 3] Strain (Grid 2)**	**[1, 2, 3] Strain (Grid 3)**	**[4, 5, 6] Strain (Grid 1)**	**[4, 5, 6] Strain (Grid 2)**	**[4, 5, 6] Strain (Grid 3)**
2	3	2	2	2	2	2
2.1	3	2	2	2	2	2
2.2	3	3	2	2	2	2
2.3	3	2	2	3	2	2
2.4	3	2	2	3	2	2
2.5	3	2	2	3	2	2
2.6	3	2	2	3	2	2
2.7	3	2	2	2	2	2
2.8	3	2	2	3	2	2
2.9	3	3	2	3	2	2
3	3	3	2	3	2	2
3.1	3	2	2	3	2	2
3.2	3	2	2	2	2	2
3.3	3	2	2	3	2	2
3.4	3	2	2	3	2	2
3.5	3	3	2	3	2	2
3.6	3	3	2	3	3	2

TABLE 3.3 Strain Values for 5 mm Diameter of the Implants at Grids

Assignment	**R1**	**R1**	**R1**	**R2**	**R2**	**R2**
Reduction method/ID	**Strain (G1)**	**Strain (G2)**	**Strain (G3)**	**Strain (G4)**	**Strain (G5)**	**Strain (G6)**
Seconds elapsed	**[1, 2, 3] Strain (Grid 1)**	**[1, 2, 3] Strain (Grid 2)**	**[1, 2, 3] Strain (Grid 3)**	**[4, 5, 6] Strain (Grid 1)**	**[4, 5, 6] Strain (Grid 2)**	**[4, 5, 6] Strain (Grid 3)**
0.1	3	4	2	3	4	3
0.2	3	4	2	3	4	3
0.3	3	4	2	3	4	3
0.4	3	4	2	3	3	3
0.5	3	4	2	3	4	3
0.6	3	4	2	3	4	3
0.7	3	4	2	3	4	3
0.8	4	4	2	3	4	3
0.9	4	4	2	4	4	3
1	4	4	2	3	4	3
1.1	4	4	2	4	4	3
1.2	4	4	2	4	4	3
1.3	4	4	2	4	4	3
1.4	4	4	2	4	4	3
1.5	4	4	2	4	4	3
1.6	4	4	2	4	4	3
1.7	4	4	2	4	4	4
1.8	4	4	2	4	4	3

TABLE 3.3 *(Continued)*

Assignment	R1	R1	R1	R2	R2	R2
Reduction method/ID	**Strain (G1)**	**Strain (G2)**	**Strain (G3)**	**Strain (G4)**	**Strain (G5)**	**Strain (G6)**
Seconds elapsed	**[1, 2, 3] Strain (Grid 1)**	**[1, 2, 3] Strain (Grid 2)**	**[1, 2, 3] Strain (Grid 3)**	**[4, 5, 6] Strain (Grid 1)**	**[4, 5, 6] Strain (Grid 2)**	**[4, 5, 6] Strain (Grid 3)**
1.9	4	4	2	4	4	4
2	4	4	2	4	4	4
2.1	4	4	2	4	4	4
2.2	4	4	2	4	4	4
2.3	4	4	2	4	4	4
2.4	4	4	2	4	4	4
2.5	4	4	2	4	4	4
2.6	4	4	2	4	4	4
2.7	4	4	2	4	4	4
2.8	4	4	2	4	4	4
2.9	4	4	2	4	4	4
3	4	4	2	4	4	4
3.1	4	4	2	4	4	4
3.2	4	4	2	4	4	4
3.3	4	4	2	4	4	4
3.4	4	4	2	4	4	4
3.5	4	4	2	4	4	4
3.6	4	4	2	4	4	4

3.3 RESULTS

Tables 3.2 and 3.3 show induced strain at each grid at regular intervals of time for implants with 4.2 mm and 5 mm when the load is applied. It shows that implants with 4.2 mm diameter induce lesser strain than the 5 mm diameter implant. The outcome of this study also shows that no significant variation is observed between the grids with respect to the implant diameter. However, due to changes observed in strain at grids, it can be seen that implant diameter influences the strain value around the bony-implant region. The calculated strain values for 4.2- and 5-mm diameter of the implant are listed in Tables 3.4 and 3.5 The strain at the mesial side shows the maximum strain value than the distal side of the bone.

TABLE 3.4 Calculated Strain Value for 4.2 mm Diameter of the Implant

Location	Maximum Principal Strain	Minimum Principal Strain
Mesial	3.207	1.793
Distal	2	2

TABLE 3.5 Calculated Strain Value for 5 mm Diameter of the Implant

Location	Maximum Principal Strain	Minimum Principal Strain
Mesial	4.081	3.707
Distal	0.9189	2.292

3.4 DISCUSSION

This experimental study was based on the SG which is used to measure certain parameters accurately on the surface of the bone block. Also, this method is considered as a non-destructive methodology which could help to measure biomechanical parameters in an effective manner [7]. However, the data calculated and shown in Tables 3.4 and 3.5 are only for the particular region. Since SG is a sensor-based methodology, it is not possible to measure the maximum strain value of the bone, which happens normally inside the bone. This is because placement of the strain gauge is near to the bone around implants. It is also observed that significant percentage reduction was observed between the strains when the two diameters of the implants are compared, and the finding of this study is in good agreement with the published work [8]. The outcome of this study indicates that implant with small diameter induced lesser strains.

3.5 CONCLUSION

This work explains the experimental methodology used to study the biomechanical characteristics of two implants with different diameter. The description of 4.2 mm and 5 mm diameter of the implants was explained, and the observed readings are noted. The dental implant diameter has significant influence on the stress/strain dispersion around the bony-implant region. Mesial and distal sides of the bone block have been used in this investigation.

KEYWORDS

- **biomechanical characteristics**
- **bony-implant region**
- **finite element analysis**
- **polyurethane foam**
- **prosthodontic treatment**
- **strain gauge**
- **universal testing machine**

REFERENCES

1. Van, S. R. C., Guan, H., & Loo, Y. C., (2006). *Comput. Methods Biomech. Biomed. Engin., 9*(4), 257.
2. Roberts, D. H., (2000). *Br. Dent. J., 128*(3), 117.
3. Salvi, G. E., & Brägger, U., (2009). *Int. J. Oral Maxillofac. Implants, 24*(9), 69.
4. Seth, S., & Kalra, P., (2013). *Int. J. Sci. Res., 2*, 121.
5. Karl, M., Taylor, T., Wichmann, M. G., & Herckmann, S. M., (2006). *Int. J. Prosthodont., 15*(1), 20.
6. Cehreli, M. C., Akkocaoglu, A., Comert, A., Tekdemir, I., & Akca, I. K., (2005). *Clin. Oral Implants Res., 16*(5), 540.
7. Vasconcellos, D. K., Özcan, M., Maziero, V. C. Â., Bottino, M. A., & Yener, E. S., (2012). *Implant Dent., 21*(3), 225.
8. Castro, C. G., Zancope, K., Verismo, C., Soares, C. J., & Neves, F. D., (2014). *Braz Oral Res., 29*(1), 1.

CHAPTER 4

DEPTH ESTIMATION USING STEREO MEDICAL IMAGING

KETKI P. KSHIRSAGAR[1] and RAJESHREE A. SHINDE (ROKADE)[2]

[1]*Electronics and Tele-Communication Department, Vishwakarma Institute of Information Technology, Pune, Maharashtra, India*

[2]*Electronics and Tele-Communication Department, Lokmanya Tilak College of Engineering, Navi Mumbai, Maharashtra, India*

ABSTRACT

Depth estimation has importance in many applications such as robot navigation, 3D movies, human-computer interfaces, medical depth detection imaging, etc. Different maps are generated using stereo communication algorithms. In-depth information is displayed using a contrast map. Generally, depth is estimated by Sum of absolute difference (SAD), Semi Global Block Matching (SGBM), and Normalized Cross Correlation algorithms. We have implemented these algorithms in the Python language. These algorithms generate a map of variance with intelligent block matching and slope limits. In experimentation, the Semi Global Block algorithm shows better results.

4.1 INTRODUCTION

Stereo vision reconstructing 3D geometric based on camera two or more perspective images. Stereo vision is like human vision, which is captured by two eyes. Stereo vision estimate distance by calibrating camera for reducing much distortion as possible. For camera calibration required to work on

Robotics and Automation in Healthcare: Advanced Applications. R. Thanigaivelan, Sanjay Singh, & Clement Christy Deepak C. (Eds.)

intrinsic and extrinsic parameters. Camera calibration is a process driven method in which properties including intrinsic and extrinsic of camera are calculated. After calibration, camera images can be utilized for the required purposes. Camera calibration is drastically needed in order to get fine-tuned parameters of object plane. Development of machine vision medical applications camera calibration is a must to get Camera efficient object parameters. A camera is a visual sensor and is very important part for various technical branches including images processing, automation, and surveillance as well entertainment sector. Camera calibration is a parameter tuning process. This involves adjusting parameters of the camera to find an exact relation between 3D coordinates and its corresponding 2D image object pixels captured by calibrated camera. Specifically, these methods or techniques used for finding parameters.

Camera lens of intrinsic parameters is radial distortion, focal length, optical center, and the lens coefficients. The camera orientation with reference coordinates included in extrinsic parameters. Most commonly used camera calibration techniques are pattern calibration, geometric clues, deep learning.

1. **Geometric Clues:** Geometric clues involve straight lines, equally distributed or vanishing points in order to calibrate.
2. **Pattern Calibration:** If taking whole set of images; the appropriate method of calibration is to capture multiple object images of known dimensions from various coordinate points. In checkerboard pattern, a circular pattern is also used for checkerboard-based method of known dimensions.
3. **Deep Learning Based:** If taking few imaging object information, Deep Learning method better to use calibration of camera.

The camera calibration determines require parameters of the object image. It's very important process in many computers vision related applications and instruments. Applications and instruments including the camera is modeled with camera internal parameters including focal length, principal lines, skew of axis and orientation of camera is explored in terms of extrinsic parameters which includes rotational and translational features. In linear or nonlinear processes, while finding internal and external camera parameters required orientation of image plane as well as pixels in the image.

Those pixels and edges are presented as a pattern calibration with known geometrical structures, generally a flat chessboard. Various studies have focused on calibrating methods, Major of them is emphasized to estimate parameters step and calibrating location factors, this calibrating problem not always quantified while automatic calibration pattern detection.

Stereo vision also estimate distance by using epipolar geometry. Epipolar geometry means stereo vison geometry. Geometric relation shows 3D image or pictures to 2D image planes. For estimating distance stereo vision also builds disparity and depth map.

Depth estimation has vital role in image processing or predict the information from images using stereo vision. We use computer vision library in python language for coding to perform algorithms. Deep scaling or subtraction refers to a set of strategies and algorithms aimed at obtaining representation of a local area structure.

Developments in three-dimensional (3D) image reconstruction and an autonomous systems application has stimulated the research on stereo depth estimation imaging. Stereo vision is bright field and vision-based applications widening science to enjoyment sectors. Driving supports for cars and unmanned aerial vehicles (UAV) are some robotic applications requiring authenticate depth estimation from camera captured optic data. Biological depth estimation is still big pending problem which is similar to artificial depth precipitation.

Stereo matching is defined as the process of obtaining the best connection points between a pair of stereo images. In addition to computer hardware limits, different map accuracy is limited by a number of operating modes such as closure, ambiguity, light variation and visual distortion. Stereo matching algorithm increases performance by using image correction and camera. Depth estimation using stereo imaging is performed in two steps: The first step is to map the difference from the stereo image binding using the matching algorithm. Inequality map contains unequal pixel between the projection of the same things in the pictures on the left and right projected images. The second step is the transfiguration of the inequality map into a depth map.

An important area of research has been the discovery of 3D event information in 2D images. Dynamic camera parameters more focus depth of vision images. But in depth from defocus, it is not possible to change the camera parameters. Tsai proposed stereo and defocus integration system. But in comparing the stereo algorithm they use rough depth measurements found in the defocus.

The generation of depth map from stereo image pair was a major problem for two decades. Major problem was Depth map estimation from stereo correspondence other problem is the estimation constraints of a stereo system, which diminish the purpose of depth maps. For enhancing the depth purpose of stereo imaging this chapter marks survey of stereo vision system framework. The process of finding the best connection points between a pair of stereo images is called stereo matching. Depth estimation from stereo vision contains two main tasks first is to estimate disparity map of stereo image pairs by means of stereo matching algorithm and second is the transforming disparity map into depth map. Fundamental Block diagram of Depth Estimation is as shown in Figure 4.1.

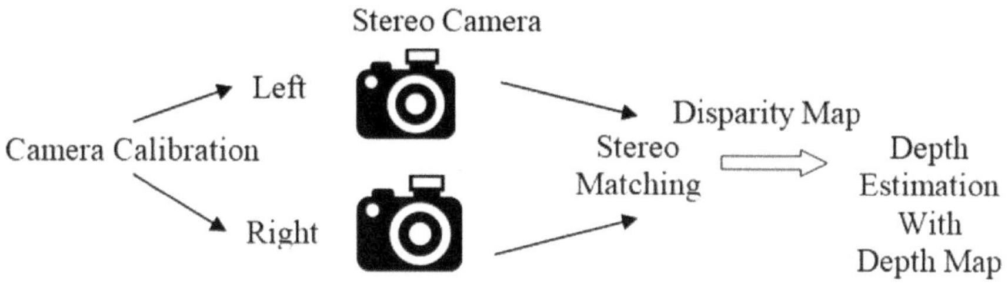

FIGURE 4.1 Fundamental block diagram of depth estimation.

Generally, SAD, SGBM, and normalized cross correlation these algorithms are used for estimating disparity maps. By using this algorithm, we compared the outputs and estimates which algorithm gives better disparity map.

4.2 LITERATURE WORK

There is an important role of depth estimation in surgical laparoscopic navigation systems. In maximum systems which been used for depth, estimation is supervised in which training data is required. But Huoling Luo et al. [1] proposed a CNN-based model in which training data is exceptional for depth estimation in laparoscopes. Huoling Luo et al. include a disparity map. The disparity map removes the wrongly predicted pixels by training of encoder and decoder network by confidence measures. It applies the neighborhood loss of smoothness factor to generate the similar pixel appearances that are near to the depth. This method improves the performance metric by improving the MAE and RMSE. This system is useful for real laparoscopic surgical navigation. An FCN-based depth estimation method proposed by Takuya Suzuki et al. [5]. In this system, a monocular camera is used to capture a laparoscopic image. Takuya Suzuki proposes a segmentation algorithm of U-Net based using color and depth information. With the MCCAI challenge, IOU, and Dice coefficient improved by 2%.

Zhen Xie et al. [4] aim to explore the problem of measuring the depth from a pair of event-based stereo sensors. In this chapter, they introduce an event-based stereo depth measurement algorithm based on message transmission. In addition to analyzing the features of a single event, the algorithm uses the Markov Random Field (MRF) to measure the barriers between adjacent events, such as disparity and continuity of depth. According to the results, this method gets more stereo similarity than other methods, and each similarity has a higher level of accuracy.

The system experimented white-light endoscopic images on depth prediction and chromo endoscopic images. A motion estimation network [6] is used for the reconstruction of the motion of the endoscopy camera. The network of depth estimation is applied on endoscopy images for calculation of depth. From video of endoscopy, without external tracking equipment Shiyuanliu et al. extracted the camera orientation and exact depth mapping. Errors are in millimeters. Endoscopy is achievable and the intuitive 3D visual assist is able to give surgeons more accurate feedback during the operation using 3D visual assist surgeons.

Monocular and Stereo Cues proposed for finding depth estimation by Ashutosh Saxena et al. [7]. They projected a hierarchical, multi-scale MRF learning model to catch single cues and integrate them into a stereo system to get advance depth measurements. Monocular signals and stereo signals provide compatible, diverse information about depth. Using monocular and stereo cues, they obtain additional correct depth measurements that are more likely to be used by monocular or stereo-only.

Satyarth Praveen [8] has proposed a powerful algorithm for calculating depth using a stereo view with other perception. To measure the distance from an object, the object is viewed at two different angles. The proposed method starts by obtaining the outstanding two input images features and then uses a mixing of multiple approaches. Based on the differentiating values, the value of the difference of the image component is coded, and the output looks much lower than normal output. The purpose of this chapter is to make the depth measurement process faster and easier by avoiding calculating the depth of unwanted pixels.

Lazaros Nalpantidis et al. [9] during this analysis chapter is predicated on robotic applications using algorithms and technologies. Robotics evolves apace and demand methods which will serve autonomous behaviors. The foremost fascinating research problems with the robotics-oriented stereo vision field is it involves the considering of week lighting provisions, the need for easy computation methods, the utilization of numerous-orienting systems, management, and also the introduction of the latest biologically galvanized methods for robotic vision.

In quantifying visual information [10], depth estimation also helps in understanding the scene. Plethora depth estimation algorithms are very well known. It is not possible sometimes to extract information from the inappropriate recognition span of an off-the-rack depth sensor and there are hard situations the depth sensor due to differences in the actual scene and setup. Therefore, the concept of endoscopic domain is completely different from the concept of depth estimation. Mostly endoscopic sequences are trickier than outdoor sequence. This system experimented and provided comparison on estimated techniques and highlighting challenges in transferring systems.

Robotic surgery is advantageous over traditional surgery. It allows surgeons to reach hard-to-reach areas through small incisions. Patients experience less pain during and after surgery. It provides advantages in dexterity. It has many more advantages like precision and 3D vision, over traditional surgery. Siamese self-supervised learning algorithm performed on stereo images is proposed by Ye et al. [11] for robotic surgery for depth estimation.

Mahmoud Abdelhamid [12] introduces a novel methodology that combines feature- and correlation-based methods to extract depth information from stereo-vision acquisitions. The proposed method is primarily implemented with a robotic manipulator in the field of autonomous pick and place. The existing vision-guided robotics suffers from a slow response time and is still relying on a priori training and teaching stages.

Huai-Mu Wang et al. [13, 14] introduce a deep convolutional neural network (CNN) based real-time object detection and depth estimation method. Specifically, they incorporate transfer connection blocks (TCBs) to enhance real-time object recognition, especially for small objects. They apply the epipolar constraint to enhance prediction accuracy and binocular vision to the monocular-based disparity estimation network for depth estimation. In order to achieve real-time detection and depth estimates, they finally merge the depth information with the two-dimensional (2D) position of the identified object. The outcomes show that the suggested strategy outperforms traditional techniques in terms of result. This approach has two parallel models: object recognition and depth mapping. Object detection is improved by using transfer connection blocks (TCBs) with CNN. For depth estimation, Monodepth was modified for binocular images. The result from this approach is better compared to a conventional method.

Multi-view stereo method and monocular endoscopic videos are used by Xingtong Li et al. [15], to supervise learning sparsely, structure from motion. For monocular endoscopy, they suggested depth estimation by self-supervised method. This system doesn't require any manual labeling and

CT scan in the coaching and petition phases for ill persons. After testing the system on unlabeled training data from the number of patients it achieves a sub-millimeter residual error.

To improve the model generalizability, a Siamese network is introduced in extension with an auto encoder and a spatial transformer. The accurate depth estimation outperforms standard stereo matching approaches conducted on Vivo videos collected during robotic surgery. A method is unsupervised in which do not require depth labels while training. It provides applicability to large-scale "Vivo video processing" where known depths are unavailable. Neck, head, urological surgeries, anorectic, head, and cardiovascular surgeries [4, 16] are done by minimally invasive surgery (MIS).

Tyler et al. [17] projected CNN algorithm for depth estimation. This chapter explores the advantage of using Convolution Neural Networks in generation of a disparity space in image for providing disparity maps from stereo imaging.

Pablo et al. [18] suggest that sometimes we cannot see certain features and characteristics in standard image acquisition systems, so we interpret information in images using depth. There are active and passive methods to process reflections, while the passive ones rely strictly on the natural energy received through the pictures. Among the latter choice, we find single cavity systems, which operate in one view, as well as a stereo or multi-view system, which operates with more than one view.

In 3D reformation and endoscopy describing, MIS is used to suppress a problem of surgical space. To suppress disproving forms near to the track of medical instruments [2, 19] MIS application is useful. The Gastro endoscopy procedure is done using a tube which is flexible and very long is called endoscope. On one end tube has a small light. The long tube is inserting into throat. Then it is pushed from throat to stomach. From stomach it is pushed into duodenum. It is difficult to generate depth images, ground truth RGB images, and computer-generated (CG) data using an in algorithm of depth estimation.

Patric Kamencay and Roman proposed improved Depth Map evaluation from Stereo image supported Hybrid model. They describe a method for retrieving depth map primarily based on hybrid segmentation algorithmic program victimization method filtering. The algorithm was tested in a really complex environment with very poor light intensity captured by an industrial stereo camera system using a bumblebee camera system.

To find out the depth of gastro endoscopy a supervised training strategy is proposed by Aji Resindra Widya et al. [20]. The first-ever 3D model by endoscopic images with indigo carmine (IC) blue dye was used for training heavy credential depth data. To use the network of depth estimation on endoscopic images without IC dye, virtual no IC images are generated from Cycle GAN.

Uma Mudenagudi et al. [21] have made a paper proposal about measuring depth using defocused stereo image pairs. In this chapter, the depth from the defocus (DFD) is combined with the stereo-based approach. In this approach, simultaneous retrieval of the depth and retrieval of the image is of great benefit. The proposed system provides better image stabilization when the image has blurrier. From this approach, the quality of the restored image is improved, and the accuracy of the depth measurement is improved.

Paul Munro et al. [22] projected Stereo Vision computer Depth Perception that the depth estimation victimization Autonomous Road Vehicle Navigation. However, provides temporary introduction of triangulation, feature primarily based disparity and area-based disparity. The most advantage of the feature-based rule is its speed. The method of finding options in each image, then scheming the disparity will simply be done in real time. The area-based method provides dense disparity maps. Other enhancements will be obtained by pre-processing the pictures before implementing the algorithm.

Baris Baykant Alagoz [23] proposed depth resolution of system that is experimented on the bases of systematic parameters of sense modality hardware systems. Experimented modality improves exploration for different efficient applications.

4.3 EXPERIMENTAL METHODS

4.3.1 SUM OF ABSOLUTE DIFFERENCE (SOAD)

Total contrast is the ratio measures similarity between two image blocks. Similarity is calculated by absolute pixel and corresponding pixel in each block and corresponding pixel block. The SAD algorithm can be defined using the following equation:

$$\mathrm{SoAD}(x,y,d) = \sum_{x,y\in w} \left(I_l(x,y) - I_r(x-d,y)\right) \qquad (1)$$

where; I_r corresponds to left image; I_r corresponds to right image (x, y) are the intensity values; and d is the disparity value.

The equation states that the algorithm considers the absolute difference between intensity of pixels the image on the right and its corresponding pixel on the left image. The differences are averaged over the window w so as to get the disparity map. The window size may be varied in step with the requirements. The SoAD algorithm is that the traditional algorithm and principally used as a result of it has terribly less computations. The algorithm is fast however the accuracy of the disparity map is a smaller amount because of noise at the object boundaries.

Figure 4.2 elaborate procedure for SoAD algorithm, we have to find the similarity between two blocks of pixels. We can find disparity between the two images. For these, take two stereo images, i.e., left, and right images and store pixel values in matrix for similarity use the sum of absolute difference method that will be used to compare pixel blocks. Pixel blocks with a minimum value for full difference are very similar to pixel blocks with a maximum value difference. Here, we compare a pixel block from the image on the left with a pixel block in the right image. Each pixel block in the picture on the left is compared to the number of pixel blocks in the picture on the right. Use the loop method to skip each component in the left image, select the pixel block in the left image, apply the comparison blocks, add the unequal values, and store value values in a matrix called the variance map or disparity map.

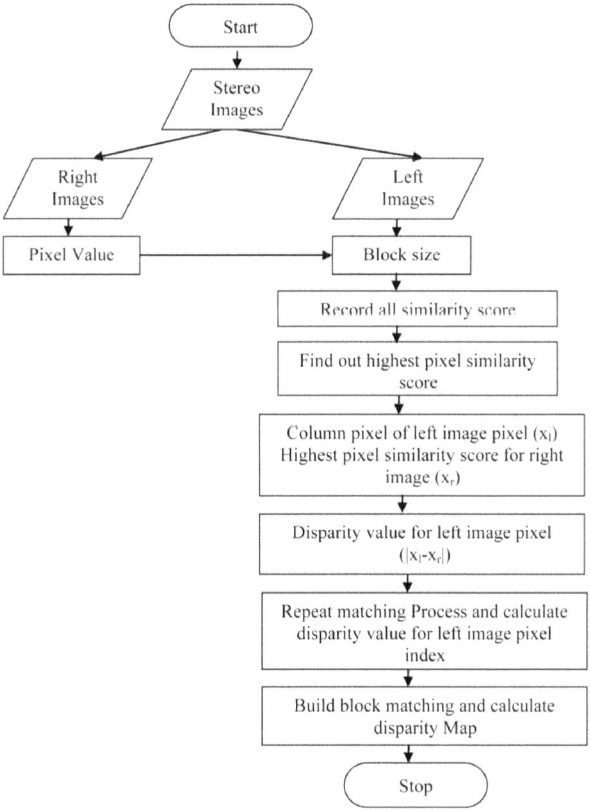

FIGURE 4.2 Sum of absolute difference.

4.3.2 *SEMI GLOBAL BLOCK MATCHING (SGBM)*

Semi global simulation (SGM) is a computer-aided viewing system that compares a map of the dense differences from the surface of a fixed stereo image. In this text the stereo algorithm is made using smart block chain simulation with a non-abstract Birchfield-Thomas metric form. It is a flexible set of systems that include searches that vary in completely different ways. Differences are then calculated in many ways, so the best match is chosen because of the very good differences. Global radiometric variation and noise are well modeled with the proposed algorithm. The algorithm is often represented as in Figure 4.3.

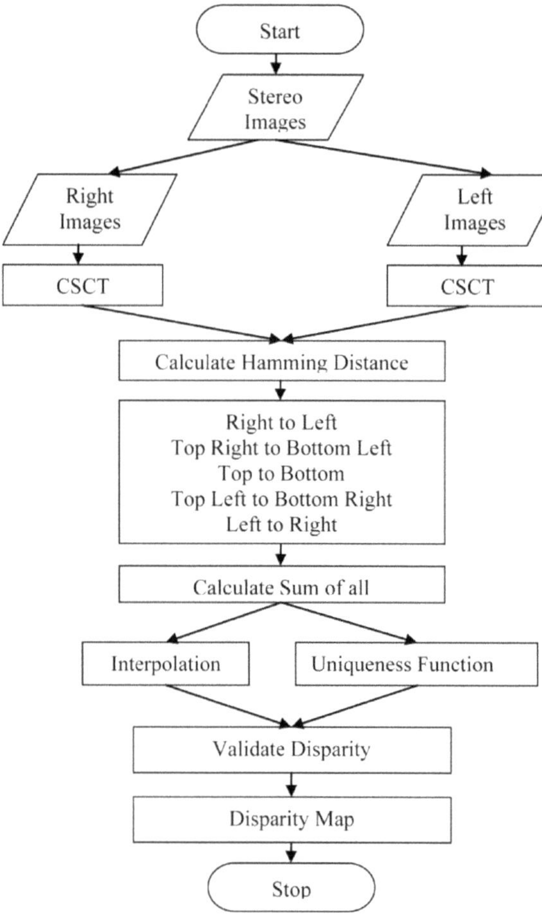

FIGURE 4.3 Flowchart of semi global block matching.

$$E(D) = \sum_{p}\left(C\left(D_p\right)\right) + \sum_{q \in N_p}\left(T\left[\left|D_p - D_q\right| = 1\right]P1 + \sum_{q \in N_p}\left(T\left[\left|D_p - D_q\right| > 1\right]P2\right.\right. \quad (2)$$

Where the first term is the sum of the combined costs of all pixel costs. The second term is that the pixel value is slightly different from its neighbor, the P1 fine is added. The third term adds a large, fixed fee of P2, for all major variations of variance. The algorithm determines the above number, which depends on the number of pixels and the variance. The low cost of small changes allows for adaptation to the sloping area. The continuous charge of large changes remains uninterrupted. However, it is necessary to make sure that the $P2 \geq P1$. The small earth block algorithm is one of all the various methods used within the stereo view. Allows closure detection, fine-tuning, and deep suspension. The proposed algorithm is powerful against radiometric contrast and image noise. The success of the SGM has encouraged many researchers in its application.

4.3.3 NORMALIZED CROSS-CORRELATION (NCC)

NCC is often used as a mark to assess the gradation of likeness (or diversity) between two comparative images. Additionally, NCC is restricted to the range of 1 and NCC, a standard interaction (standard connection), which is used to create a level of correlation between the intended targets. Original image pixel coordinates (p_x, p_y). Build a n×n is matching aperture. Then attack position pixel (p_x+d, p_y), also build a n×n the size of the matching aperture is measured the similarity between two images, where d has a value range. For measuring NCC, first two image frames are corrected on similar base on horizontal position and its optical center and epipolar line also on horizontal. When matching process executed in the inclined epipolar direction it consumes more computational resources, so avoid this situation. Normalized cross correlation algorithm shown in Figure 4.4.

$$NCC\left(I_1, I_2\right) = \frac{\sum_{x}(I_1(x) - \mu_1)(I_2(x) - \mu_2)}{\sqrt{\sum_{x}(I_1(x) - \mu_1)^2 (I_2(x) - \mu_2)^2}} \quad (3)$$

where; $I_1(x)$ is the pixel value of the original image; μ_1 is the mean value of pixels in the original window; $I_2(x)$ is the total pixel value for first image in corresponding point on the target image offset by ‘d’ in the ‘x’ direction; and μ_2 match the window pixel average value for the target image.

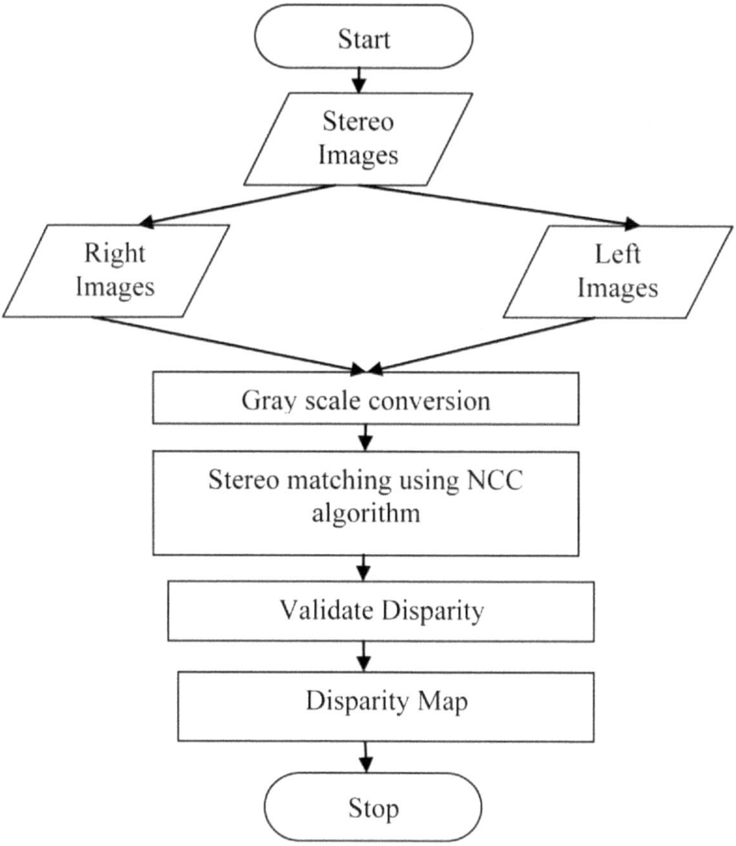

FIGURE 4.4 Flow chart of normalized cross-correlation.

4.4 CONCLUSION

Generally, SGBM algorithm, SoAD, and NCC used for calculating disparity map for depth estimation for any application like robot navigation, 3D movies, Human computer interfaces, medical depth detection imaging, etc. SGBM one of algorithm shows betterment for finding depth estimation for many medical imaging applications.

KEYWORDS

- **absolute difference**
- **aerial vehicles**
- **cross-correlation**
- **epipolar line**
- **medical imaging applications**
- **semi-global block-matching**
- **three-dimension**

REFERENCES

1. Huoling, L., Qingmao, H., & Fucang, J., (2019). Details preserved unsupervised depth estimation by fusing traditional stereo knowledge from laparoscopic images. *Healthcare Technology Letters*, 154–158.
2. Kim, J., Al Faruque, H., Kim, S., Kim, E., & Hwang, J. Y., (2019). Multimodal endoscopic system based on multispectral and photometric stereo imaging and analysis. *Biomed. Opt. Express, 10*(5), 2289–2302.
3. Mahmoud, N., Collins, T., Hostettler, A., Soler, L., Doignon, C., & Montiel, J. M. M., (2019). Live tracking and dense reconstruction for handheld monocular endoscopy. *IEEE Trans. Med. Imaging, 38*(1), 79–89.
4. Zhen, X., Shengyong, C., & Garrick, O., (2017). Event-based stereo depth estimation using belief propagation. *Frontiers in Neuroscience, a Section of the Journal Neuromorphic Engineering,* 1–14.
5. Takuya, S., Keisuke, D., & Yoshito, M., (2019). Depth estimation for instrument segmentation from a single laparoscopic video toward laparoscopic surgery support. *International Conference on Intelligent Medicine and Image Processing* (pp. 21–24).
6. Shiyuan, L., Jingfan, F., Dengpan, S., Tianyu, F., Yucong, L., Deqiang, X., Hong, S., et al., (2022). Joint estimation of depth and motion from a monocular endoscopy image sequence using a multi-loss rebalancing network. *Biomedical Optics Express, 13*(5), 2707–2727.
7. Ashutosh, S., Jamie, S., & Andrew, Y. N., (2007). Depth estimation using monocular and stereo cues. *Proceedings of the 20th International Joint Conference on Artificial Intelligence* (pp. 1–7). Hyderabad, India.
8. Satyarth, P., (2019). Efficient depth estimation using sparse stereovision with other perception techniques. *Advanced Image and Video Coding* (pp. 1–20).
9. Lazaros, N., & Antonios, G., (2011). Stereo vision depth estimation methods for robotic applications; Chapter 21. In: *Book: Depth Map and 3D Imaging Applications: Algorithms and Technologies* (pp. 397–417).

10. Lalith, S., Lukas, B., Georgii, K., Ivo, W., Matthias, K., Raffaele De, S., & Sandy, E., (2020). Domain gap in adapting self-supervised depth estimation methods for stereo-endoscopy. *Biomedical Engineering, 6*(1).
11. Ye, M., Johns, E., Handa, A., Zhang, L., Pratt, P., & Yang, G. Z., (2017). *Self-Supervised Siamese Learning on Stereo Image Pairs for Depth Estimation in Robotic Surgery*. arXiv preprint arXiv:1705.08260.
12. Mahmoud, A., (2011). *Extracting Depth Information from Stereo Vision System, Using a Correlation and a Feature Based Methods* (pp. 1–92). Clemson University, Tiger Prints.
13. Huai-Mu, W., Huei-Yung, L., & Chin-Chen, C., (2020). Object detection and depth estimation approach based on deep convolutional neural networks. *IEEE International Conference, on Systems, Man, and Cybernetics (SMC)* (pp. 1–17). Toronto, ON, Canada.
14. Ketki, P. K., & Dharmpal, D., (2010). Object-based key frame selection for hand gesture recognition. In: *2010 International Conference on Advances in Recent Technologies in Communication and Computing* (pp. 181–185).
15. Xingtong, L., Ayushi, S., Masaru, I., & Gregory, D. H., (2020). Dense depth estimation in monocular endoscopy with self-supervised learning methods. *IEEE Transactions on Medical Imaging, 39*(5), 1438–1447.
16. Li, C., Gu, X., Xiao, X., Lim, C. M., & Ren, H., (2019). A robotic system with multichannel flexible parallel manipulators for single port access surgery. *IEEE Trans. Indust. Inform., 15*, 1678–1687.
17. Tyler, S. J., & Skanda, S., (2015). *Usings CNNs to Estimate Depth from Stero Imagery* (pp. 1–5). Joint assignment with Jayant Thatte CS 229.stanford.edu.
18. Pablo, R. S., Belén, R. M., & José, M. S. P., (2012). *Depth Estimation an Introduction* (pp. 93–118). Carlos University of Madrid and the Spanish center for subtitling and audio description (CESyA), Chapter 5.
19. Chu, Y., Li, X., Yang, X., Ai, D., Huang, Y., Song, H., Jiang, Y., et al., (2018). Perception enhancement using importance-driven hybrid rendering for augmented reality based endoscopic surgical navigation. *Biomed. Opt. Express, 9*(11), 5205–5226.
20. Aji, R. W., Yusuke, M., Masatoshi, O., Sho, S., Takuji, G., & Kenji, M., (2021). Self-supervised monocular depth estimation in gastroendoscopy using GAN-augmented images. *SPIE 11596, Medical Imaging 2021: Image Processing.*
21. Uma, M., & Subhasis, C., (1999). Depth estimation using defocused stereo image pairs. Electronics and telecommunication department. In: 7th *IEEE International Conference on Computer Vision* (pp. 1–6).
22. Paul, M., & Antony, P. G., (2016). *Stereo Vision Computer Depth Perception* (pp. 1–18). Computer science Massey university, Albany New Zealand.
23. Baris, B. A., (2016). A note on depth estimation from stereo imaging systems. Anatolian science. *Bilgisayar. Bilimleri. Dergisi., 1*(1), 1–6.

CHAPTER 5

DEVELOPMENT OF DRONES FOR HEALTHCARE APPLICATIONS

S. A. MENON RAJKUMAR[1], S. KALAIBHARATH[1], T. NIRANJAN[2], A. GOKUL[1], and R. THANIGAIVELAN[1]

[1]*Department of Mechanical Engineering, Muthayammal Engineering College, Rasipuram, Namakkal, Tamil Nadu, India*

[2]*Department of Mechanical Engineering, Mahatma Gandhi Institute of Technology, Hyderabad, Telangana, India*

ABSTRACT

The 3D printed drone is created, and after several components are successfully put together and programmed to enable flight, the drone is launched. Further research was necessary to examine the drone's ability to carry payloads and its piloting capabilities. The drone that uses 3D printing reduces the cost of raw materials, payloads, and spare parts printing when needed.

5.1 INTRODUCTION

In the current scenario, the remote transportation of medical products, namely blood samples, medical equipment, vaccines, test kits, and medical aids, is required. Before the pandemic, the general mode of transportation is by ambulances, cars, airplanes, and helicopters. Worldwide, with rapid advancement in technology, replace the conventional mode of transportation with drones or unmanned aerial vehicles (UAVs) to enhance mobility over difficult geographical barriers, reduce carbon emissions, and also improve cost-effectiveness of healthcare delivery [1].

Robotics and Automation in Healthcare: Advanced Applications. R. Thanigaivelan, Sanjay Singh, & Clement Christy Deepak C. (Eds.)

Drones were classified into four major types, namely [2]:

- Multi-rotor drones;
- Fixed wing drones;
- Single rotor helicopter;
- Fixed wing hybrid VTOL.

Multi-rotor drones are classified based on how many motors are attached to it:

- Tricopter (3 rotors);
- Quadcopter (4 rotors);
- Hexacopter (6 rotors);
- Octocopter (8 rotors).

Among these four Quadcopter is found to be suitable for its advantages such as good lifting capacity and better stability. In this type, clockwise and counterclockwise propeller cancels the effect of turbulence. In these four motors are positioned at 90' to each other forming a '+' or 'cross' configuration. Recently 'H and X' configuration is also considered for photography application and '+' configuration is used for flying [2] (Figure 5.1).

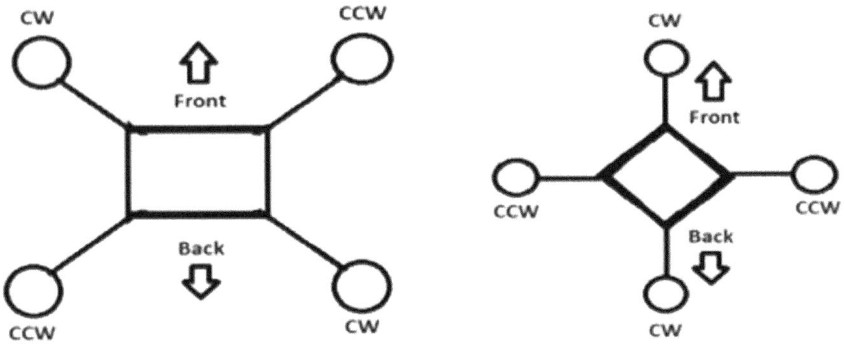

FIGURE 5.1 Quadcopter with cross and + configuration.

5.1.1 3D PRINTING OF DRONES

The need for 3D printed drones arises due to drones colliding and damage from high altitude. To overcome this issue, drone developers are now looking for new avenues towards 3D printing of frames and accessories for easy drone repairs. These frames contend in the same class as purpose-built drones, and hence must be capable to accomplish similar strength and speed requirements.

The 3D printed frames are cheap and survive a mild crash while preserving its ability to fly. Various materials such as PLA, ABS, ABS-PC, Carbon fiber are used for 3D printing. Meenakshipriya et al. [3] have designed and developed a Quad Drone using Solid Works and fabricated through 3D printing technology. They considered PLA material for black body (poly-lactic acid) and ABS material for drone landing white box (acrylonitrile butadiene styrene). Ultimaker and CubePro 3D printers were used for the purpose of producing parts with high precision dimension and good surface roughness. The electronic components such as flight control, speed controllers (SCs), transmitter and receiver with six channels, 2,280 KV brushless DC motors, propellers, and BMP180 sensor for measuring altitude (Pressure type). The challenges encountered during the drone development were errors during assembly, sizing, and weight balancing. Rangel [4] have developed a low-cost 3D printed drone system for pandemic scenario. They developed Cargo Drone to carry medical equipment and/or medicines, second Aerial Eye Drone, to check the people temperature on the ground using thermal cameras and third system, is called by Sterilizer Drone for spraying/disinfect the environment. The use of 3D printer allowed us to increase the automation of the manufacturing process. The Drone operation will be observed in real time by the pilot using IoT devices. Hiebert et al. [5] have reviewed the application of drone in healthcare. Their overall, findings specifies that drones may correspond to an economically feasible means to support healthcare for those in difficult areas-to-reach. Moreover, additional work is necessary to completely comprehend the costs to healthcare institution and the communities they serve. Purahong et al. [6] have developed the application to deliver automated external defibrillator (AED) during emergency. This application was installed on the mobile phone or tablet of the patient or the accompanying person. During the time of heart attack, the patient or the accompanying person can seek an assistance of medical drone by sending coordinates and a notification to medical staff. This application is successful during simulation studies, the operational field test obtained satisfactory results. The medical drone can be able to land within 1.5 m of the target and it can be used for transporting first aid kits and drugs suitable for medical care. Mohamed Zain et al. [2] have presented the design of an Unmanned Aerial Vehicle using the 3DEXPERIENCE software. The designing of frame parts follows the safety norms and industry standards. The design steps start with selection of materials namely polylactic acid, acrylonitrile styrene acrylate, and ABS. The drone frame consists of four parts, namely center top cover, side top cover, middle cover, and the drone's arm of weight 50 g, 10 g, 30 g, and 80 g, respectively. Habsi et al. [8] have designed and implemented an indoor autonomous flight control for drone. Autonomous flight is attained when the drone is capable of following a pre-defined path without any human intervention. A control algorithm for drone was designed by Mathworks under software MATLAB\Simulink, additionally a Vicon capture system consisting of a set of cameras was incorporated in real time with the drone MATLAB\ Simulink. Vicon capture system capture the position of drone and compare it with the estimated position, which is closer to the real position than the sensors estimation. Ashok et al. [9] have developed a prototype of drone to save human lives by increasing the delivery speed at topmost working hours. Quadcopter technology is used to distribute blood to the hospitals from blook banks and also distribute from hospital to another. Global Positioning System (GPS) is used along with android application for tracking this quad copter for aero ambulance. This type of drone can reach the destination in spite of the high traffic and altitude.

Elsayed et al. [10] have presented a novel application of an autonomous Drone for detecting of medical masks using deep learning to categorize people based on their mask-wearing with high accuracy through a classifier realized based on MobileNetV2 architecture. The training was done on TensorFlow, OpenCV, and Keras. This Drone is controlled by mobile app with help of IoT technology namely the TeamViewer app, which controls the mobile, and the QGroundControl app through the MAV-link protocol. Indra et al. [7] have developed a guided drone energized by Radio Frequency Energy Harvesting. A 32-bit Acro Naze32 Brushed Flight Controller programmed with BetaFlight software operating at 3.7-Volts is used for this development. A dedicated 3-Watt power transmitter with frequency of 915 MHz were considered and this drone was able to fly in indoor and outdoor environment, take off from ground and hover. This drone covers the maximum range of 200 meter, while the battery cycle of the constructed drone is 12 minutes. Hence it is evident that drones are applied for medical application and most of the authors have reviewed the application of drone in medical field. In this chapter, the development of drone using flexible 3D printed frame is reported which finds application for transporting the medical kits to the patients suffering from airborne disease.

5.2 DESIGN AND DEVELOPMENT OF DRONE PARTS

The drone components are drawn using solid works the different parts namely right and left leg on the back, top, and bottom portion of the drone body and right and left leg on the front as depicted in Figures 5.2–5.7.

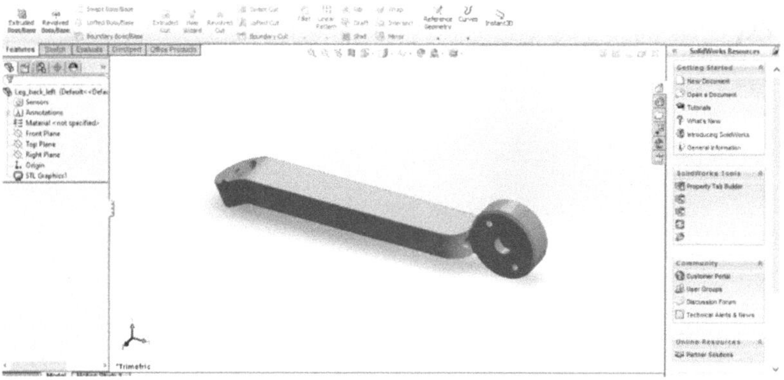

FIGURE 5.2 3D screen shot of left leg at the back side of the drone.

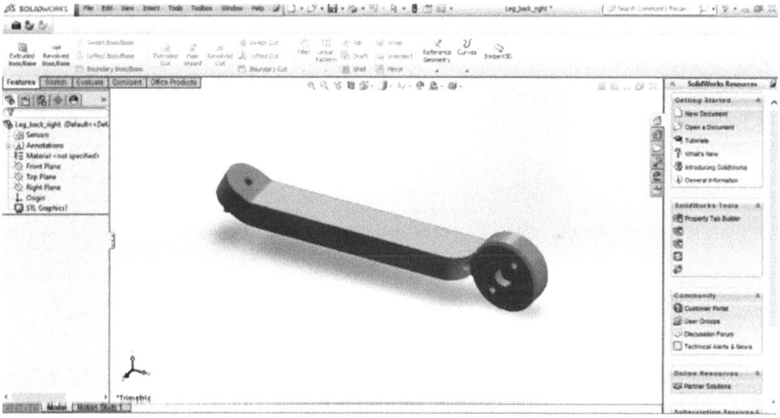

FIGURE 5.3 3D screen shot of right leg at the back side of the drone.

FIGURE 5.4 3D screen shot of middle portion of the drone.

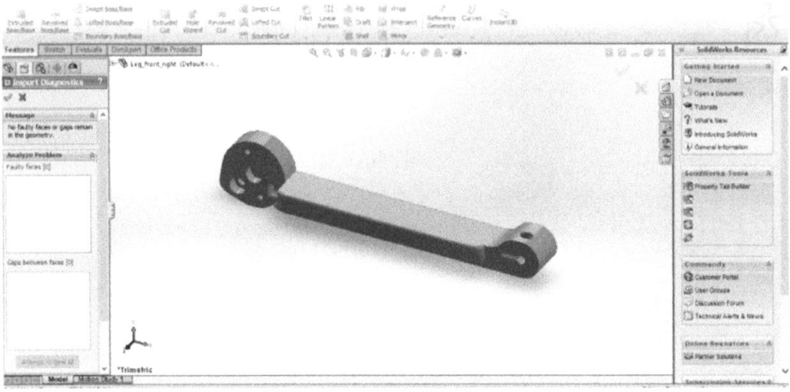

FIGURE 5.5 3D screen shot of front right leg of the drone.

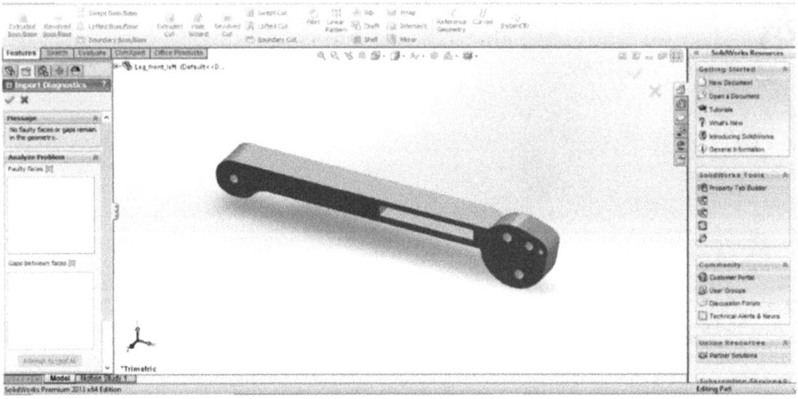

FIGURE 5.6 3D screen shot of front left leg of the drone.

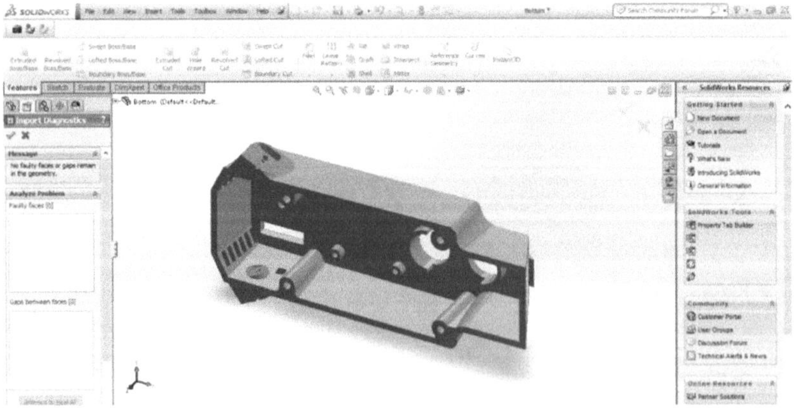

FIGURE 5.7 3D screen shot of bottom portion of the drone.

5.2.1 3D PRINTING OF COMPONENTS

Figure 5.8 shows the 3D printer used for the drone parts development. It is a fused deposition modeling (FDM) machine with maximum build volume of 140 L × 150 W × 140 H mm and layer resolution is 50–400 microns. The filament diameter is 1.75 mm with a nozzle diameter of 0.4 mm. The software named FlashPrint compatible file types are STL, OBJ, 3MF with control panel of 3.5-inch full color touch screen. The system with window and macOS is considered for the development. The printer dimension is 420 x 420 x 570 mm and has USB cable and Stick Wi-Fi connectivity.

FIGURE 5.8 3D printing machine.

Figures 5.9–5.11 show the development of model using flashforge dreamer software.

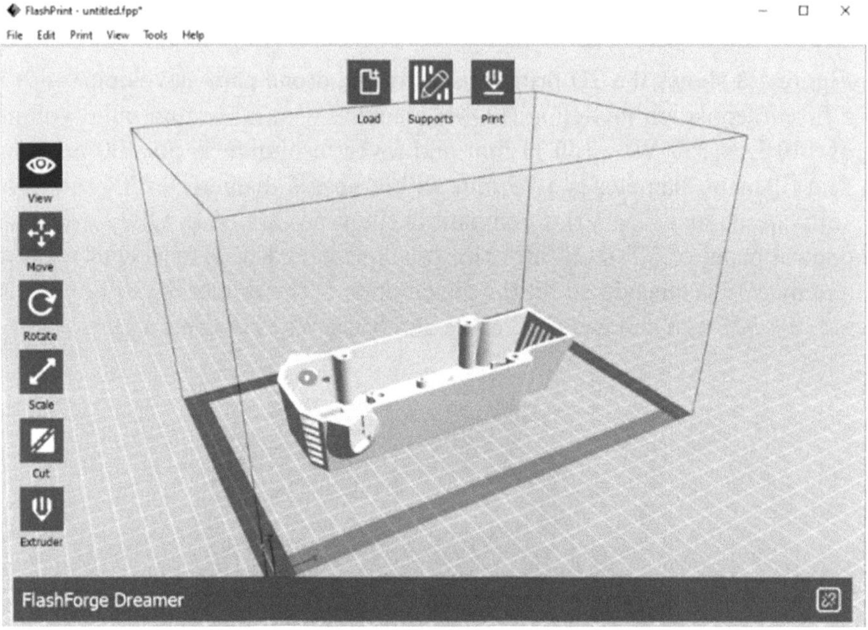

FIGURE 5.9 The center body of the drone in software.

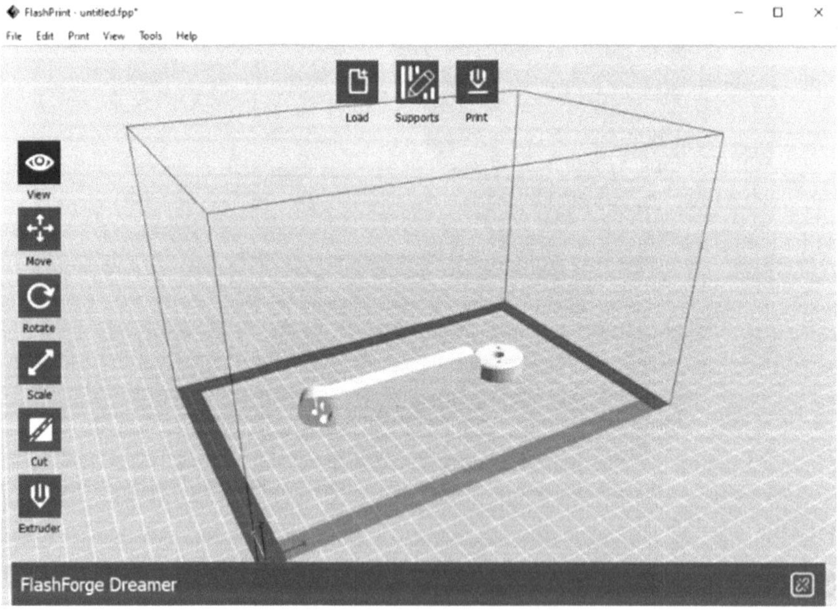

FIGURE 5.10 The side arm of the drone in software.

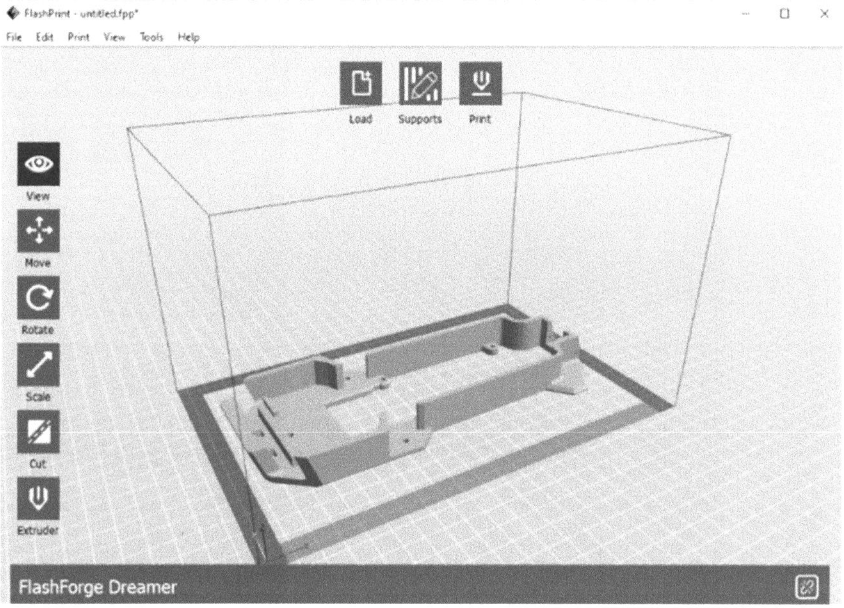

FIGURE 5.11 The top portion of the center body of the drone in software.

5.2.2 3D PRINTED COMPONENTS

5.2.2.1 FRAME

The main body of the drone is called frame. Drone frame is designed with "X"-style along with 4 arms outspreading from body as shown in Figure 5.12. Other components such as battery, camera setup, rotors, etc., were placed in frame.

5.2.2.2 PROPELLERS AND MOTOR

Propellers also called props, rotors, or blades come in various shapes, sizes, and materials. Props have the basic purpose to get your drone into the air. Figure 5.13 shows the propeller and motor used for the drone development. The motor assembly is responsible for spinning your props and providing enough thrust to fly.

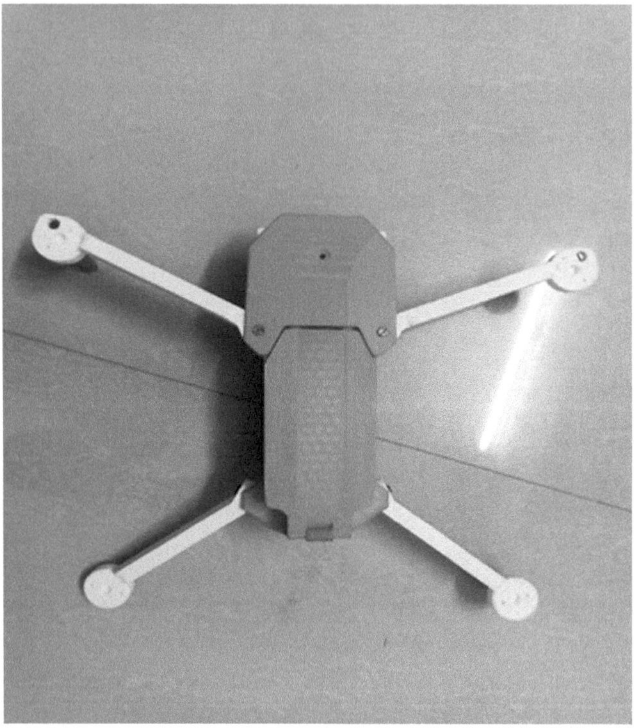

FIGURE 5.12 Frame of the drone.

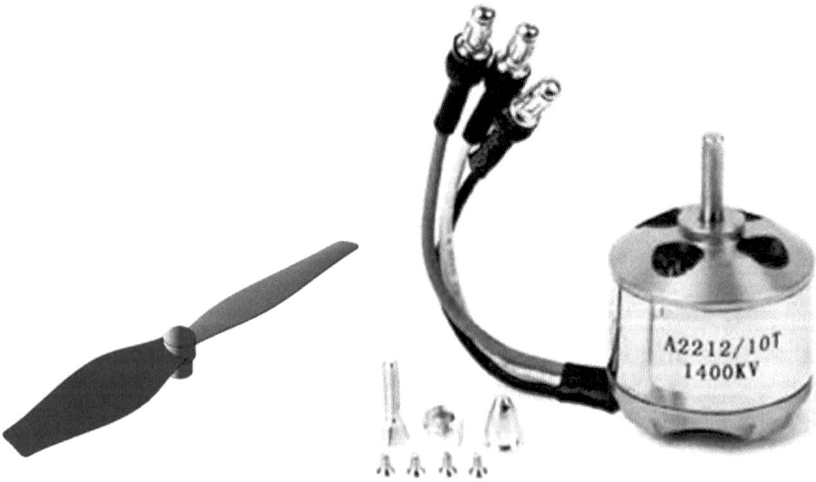

FIGURE 5.13 Propellor and brushless motor.

5.2.2.3 BATTERY

Drone battery is the heart of the drone which is done with Li-ion polymer with 2, 3, or 4S configuration. Lithium polymer (LiPo) batteries offer the best combination of energy density, power density, and lifetime on the market (Figure 5.14).

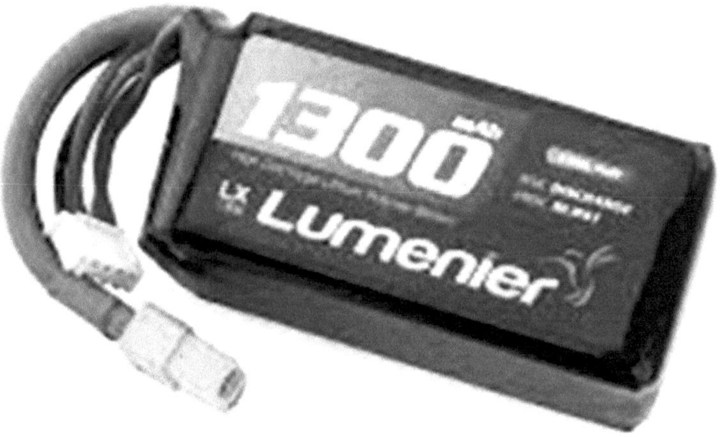

FIGURE 5.14 Battery.

5.2.2.4 FLIGHT CONTROLLER AND ELECTRONIC SPEED CONTROLLER

If the frame is skeleton, and the battery is a heart, flight controller is a brain of the drone. The flight controller understands the signals and translates those signals into action, whether that's speed and steering adjustment, camera ON/OFF, etc. The Electronic speed controller (ESC) is another vital component that controls the speed of electronic motors. Every motor has specific ESC connection which means separate ESCs will be installed on each arm (Figure 5.15).

5.2.2.5 RECEIVER AND CAMERA

Receivers are similar to sensory nerves which collect information (signals) from external sources (the radio transmitter) and transmit them to the brain (flight controller). Normally, they use an integrated antenna, which receives radio signals. The receiver has four channels for control: pitch, roll, throttle, and yaw. On the other hand, numerous receivers have supporting channels for performing definite functions or triggering flight modes. The developed drone is for medical application hence the endowment for aerial view, recording cinematography, or perspective in first-person view (FPV) flying is provided. It includes additional components namely gimbals, transmitters, signal transmitting antennas, and goggles. Go Pro or other compact high-definition video unit with onboard storage can also be considered.

FIGURE 5.15 Flight controller and ESC.

5.3 STEPS TO BUILD A DRONE

- **Step 1:** Collect all the parts required to build the drone.
- **Step 2:** Create A plan of action and make sure every component is tested.
- **Step 3:** Assemble the frame.
- **Step 4:** Fixing motors in frame.
- **Step 5:** Programmed the flight controller.
- **Step 6:** Attach the flight controller on the frame.
- **Step 7:** Attach the propellers and propeller guard on the motors.
- **Step 8:** Attach motors to the flight controller.
- **Step 9:** Calibrate and trim the drone.
- **Step 10:** Flight test.

5.3.1 WORKING PRINCIPLE

Figure 5.16 show the 3D printed drone and to achieve controllable/autonomous flight, the drone mainly needs the attitude control, shooting/measuring information, storage/transmission of information, and environmental welfare (specifically for anti-collision). An initial point of the automatic control theory is the closed-loop feedback control, i.e., by adjusting the applied input, the variation in the output is noted and response is varied to input up to the control value attained the target value. For example, the real time data is measured and any deviation in the track is sensed, and equivalent yaw adjustment is done. Environmental welfare is the application of several sensors such as optical cameras, ultrasound, etc., to sense the pathway blocks on the drone path, namely buildings, bridges, etc., to make maneuver without collision. The drone moves up and down using the rotor and relative force due to rotor pushes and pull effect makes the move of drone up and down. The fast rotor movement induces high lift. Rightward movement was achieved when reduction in angular velocity was achieved. But unbalanced thrust from rotor can result in change of direction of the drone, but the upward force generated by rotor is not same as the gravity level, hence the drone drops.

FIGURE 5.16 3D printed drone.

5.3.2 CALCULATION OF THE THRUST FORCE FOR MOTOR AND PROPELLER

Pinch test is the test conducted on many propellers which gives, thrust on each propeller. The main parameter affecting the thrust force:

- Propeller (pitch, diameter, material);
- Motor (RPM);
- Atmosphere (air).

The two propeller of size 10 × 4.5 inch, i.e., diameter of propeller is 10 inches with 4.5-inch pitch. According to the theory the diameter is related to thrust force and if the diameter increases the thrust force increases. Also, the pitch is related to the speed of the drone, if the pitch is increased the speed will be increased at the same the increase in pitch causes the turbulence in the drone. Totally 3 batteries were used with 4.7 V each and total voltage is maximum of 11.1 Volt in this drone and selection of battery is the important task. Maximum Ampere rating is calculated by multiplying maximum current and number of motors.

5.3.3 FLIGHT TIME CALCULATION

$$\text{Flight Time} = \frac{\textit{battery capacity in Ampere hours}}{\textit{average ampere drawn}} \times 6$$

Normally, the drone is landed at 20% of battery capacity:

$$\text{Average flight time} = \text{Flight time} \times 0.8$$

5.3.4 PROGRAMMING OF A DRONE

A drone developer needs to think about the logic for a particular application namely to transport the medicine to the patients in the hospital. The drone gets the data about the orientation through sensors. These values obtained from sensors makes the drones perform tasks, as the drone is also a machine and programming help to instruct it. When the drone is tossed up in the air it will go up for a fraction of second and then start falling downwards. As soon as it leaves the hand, the force that is acting on the drone is gravity and hence the accelerometer reading will show zero in vertical axis. Thus, logic would be that when the accelerometer gives zero, drone should arm itself. Flow chart given in Figure 5.17 will helps to understand the entire logic in a better way.

5.3.5 CODING OF DRONE

To simplify the flying experience, AUTO TAKE OFF feature can be added into a drone. A drone is said to have taken off successfully if it reaches a satisfactory vertical height. Hence, the important aspects of this feature would be to arm the drone and let the drone reach a satisfactory altitude above the ground (Figure 5.18).

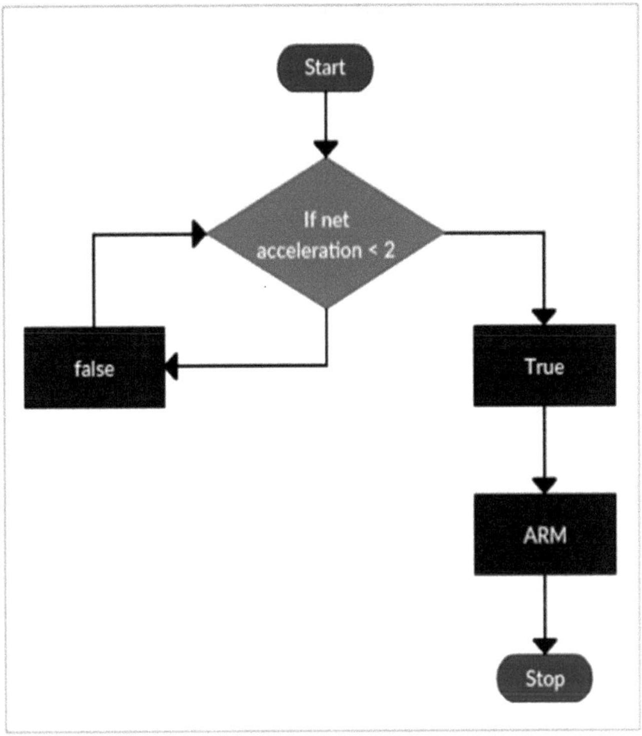

FIGURE 5.17 Flow chart.

DEPENDENCIES

```
############DEPENDENCIES####################

from drone kit import connect,

 Vehicle Mode, Location Global Relative, API Exception

import time

import socket

import exceptions

import math
```

FUNCTIONS

```
##Function to arm the drone props and takeoff at target Height (m)

Defarm _ and _takeoff(targetHeight):

        while vehicle is_armable!=True:

                print("Waiting for vehicle to become arm able.")

                time sleep(1)

        print("Vehicle is now arm able")

        vehicle mode = Vehicle Mode("GUIDED")

        while vehicle mode!='GUIDED':

                print("Waiting for drone to enter GUIDED flight mode")

                time sleep(1)

        print("Vehicle now in GUIDED MODE. Have fun!!")

        vehicle armed =True

        while vehicle armed==False:

                print("Waiting for vehicle to become armed.")

                time sleep(1)

        print("Look out! Virtual props are spinning!!")

        vehicle simple _takeoff(targetHeight)

        while True:
```

MAIN EXECUTABLE

```
####sim_vehicle py opens up port on localhost 14550

vehicle = connect('127.0.0.1:14550',wait_ready=True)

####Arm the drone and takeoff into the air at 5 meters

Arm _and _takeoff(5)

print("Vehicle reached target altitude")

####Once drone reaches target altitude, change mode to LAND

Vehicle mode=Vehicle Mode('LAND')

while vehicle mode!='LAND':

        print("Waiting for drone to enter LAND mode")

        time sleep(1)

print("Vehicle now in LAND mode. Will touch ground shortly.")
```

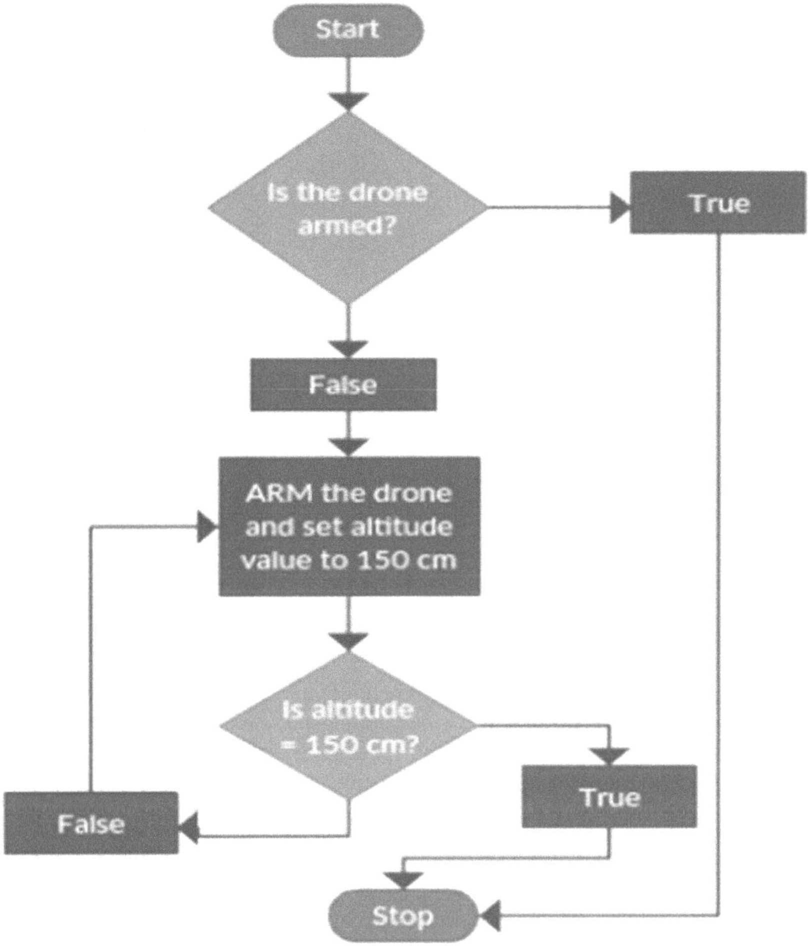

FIGURE 5.18 Program execution chart.

5.4 CONCLUSION

The 3D printed drone is developed and various parts are successfully assembled, programmed to attain the flight of the drone. More studies were to be involved to study the payload carrying capacity and piloting of the drone. The 3D printed drone cut down raw material costs, payloads, and can print spares whenever required.

KEYWORDS

- **automated external defibrillator**
- **fused deposition modeling**
- **global positioning system**
- **lithium polymer**
- **speed controllers**
- **unmanned aerial vehicle**

REFERENCES

1. Damoah, I. S., Ayakwah, A., & Tingbani, I., (2021). Artificial intelligence (AI)-enhanced medical drones in the healthcare supply chain (HSC) for sustainability development: A case study. *Journal of Cleaner Production, 328*, 129598.
2. Mohamed, Z. A. O., Chua, H., Yap, K., Uthayasurian, P., & Jiehan, T., (2022). Novel drone design using an optimization software with 3D model, simulation, and fabrication in drone systems research. *Drones, 6*(4), 97.
3. Meenakshipriya, B., Raja, N. B., Faroshkhan, S., & Matris, S., (2021). Design and fabrication of 3D printed quad drone for altitude measurement. *International Journal of Aerospace System Science and Engineering, 1*(1), 85–94.
4. Rangel, R. K., (2021). Development of low-cost medical drone, using COTS equipment. In: *2021 IEEE Aerospace Conference (50100)* (pp. 1–12). IEEE.
5. Hiebert, B., Nouvet, E., Jeyabalan, V., & Donelle, L., (2020). The application of drones in healthcare and health-related services in North America: A scoping review. *Drones, 4*(3), 30.
6. Purahong, B., Anuwongpinit, T., Juhong, A., Kanjanasurat, I., & Pintaviooj, C., (2022). Medical drone managing system for automated external defibrillator delivery service. *Drones, 6*(4), 93.
7. Indra, W. A., Industrial, S. G. H., Zamzam, N. S., Bin, H. F. N., & Zuska, F., (2021). Development of a guided drone powered by radio frequency energy harvesting. In: *2021 IEEE International Conference in Power Engineering Application (ICPEA)* (pp. 127–131). IEEE.
8. Al Habsi, S., Shehada, M., Abdoon, M., Mashood, A., & Noura, H., (2015). Integration of a Vicon camera system for indoor flight of a parrot AR drone. In: *2015 10th International Symposium on Mechatronics and its Applications (ISMA)* (pp. 1–6). IEEE.
9. Ashok, K. R., Arulselvan, P., Ashif, A., Gokul, S., & Kuppusamy, R., (2019). Aero ambulance quad copter-based technology for an emergency healthcare. In: *2019 5th International Conference on Advanced Computing and Communication Systems (ICACCS)* (pp. 1197–1200). IEEE.

10. Elsayed, E. K., Alsayed, A. M., Salama, O. M., Alnour, A. M., & Mohammed, H. A., (2021). Deep learning for COVID-19 facemask detection using autonomous drone based on IoT. In: *2020 International Conference on Computer, Control, Electrical, and Electronics Engineering (ICCCEEE)* (pp. 1–5). IEEE.

CHAPTER 6

MODELING AND OPTIMIZATION OF 3D PRINTING PROCESSES: A LITERATURE REVIEW

JYOTISMAN BORAH[1] and M. CHANDRASEKARAN[2]

[1]*Research Scholar, Department of Mechanical Engineering, NERIST, Arunachal Pradesh, India*

[2]*Professor, Department of Mechanical Engineering, NERIST, Arunachal Pradesh, India*

ABSTRACT

Three-dimensional printing (3D) or 3D printing is a recent technique used for the production of high precision-rapid tooling using high load metal components. The process is one of the subsets of additive manufacturing that produces solid objects from digital model data without loss of any material; it produces parts having a complicated shape and complex inner structure with a very high degree of dimensional accuracy. Fused deposition modeling (FDM), stereolithography (SL), selective laser sintering (SLS), and selective laser melting (SLM) are commonly used 3D printing methods. Materials such as polylactic acid (PLA), nylon, polyethylene terephthalate glycol (PETG), acrylonitrile butadiene styrene (ABS), polyether ether ketone (PEEK), and metallic powders of low-carbon steel, stainless steel, copper, and super alloys (Ti-6Al-4V) are used for printing. Considering uncertainty in the manufacturing process, the development of predictive modeling and parametric optimization of various performance measures found important for economic production of quality components with a sustainable approach.

Robotics and Automation in Healthcare: Advanced Applications. R. Thanigaivelan, Sanjay Singh, & Clement Christy Deepak C. (Eds.)

This manuscript reviews various works published in the area of 3D printing and discusses issues and future research directions.

6.1 INTRODUCTION

Three-dimensional (3D) printing is the process of development of rapid tooling which involves deposition of material in layer-by-layer form till the desired size of the product is achieved using a CAD model. It is an alternative to traditional subtractive manufacturing technique in which the desired shape and size is obtained from a basic component by performing various material removal processes. The applications of 3D print parts are commonly used in medical field, construction field, and various manufacturing applications. 3D printing technology is first commercialized in the mid-1980s, and it is widely used in automotive and biomedical sector for designing intricate/complex components relatively with ease. There are more than 30 different manufacturing methods in the production of 3D parts, involving direct and indirect approaches. A number of methods work on similar principle when observed sensitively but in details they differ considerably. The different methods of 3D technologies differ in terms of temperature and method at which the raw material power sintered. The popular 3D printing technologies include direct metal laser sintering (DMLS), selective laser melting/sintering (SLM/SLS), FDM, etc. In this chapter, the various process parameters for various 3D printing methods and their influence on various responses will be discussed.

6.2 3D PRINTING METHODOLOGY

The manufacturing process involves forming of thin layers of cross section successively (depositing layer over layer) until the final shape is achieved. Stereolithography (STL) procedure is very common; in which a point laser with several mirrors (called galvanometers) focused towards on the object to be printed which then cures and solidifies as 3D part. Figure 6.1 presents different processing steps of 3D printing.

The methodology initially designs CAD model of the component using different software packages (solid works, open SCAD, etc.). Open SCAD is commonly employed for availability and superiority in terms of easy transformations, availability, and programming ability. The designed model gives information about the 3D structure and various parameters. During compilation CAD model is represented in intermediate STL format and generate surface mesh in 3D coordinate system; and then CAD model is sliced to obtain G-codes so as to obtain 3D parts. The path to be followed by the print head, i.e., the tool path is determined by the slicing done in the STL format. The printer controls low level hardware devises such as heating element, motor operation and cooling fan so that the material is melted and extruded through nozzle to build the product layer by layer to complete the required 3D shape. Post processing involves finishing of parts by trimming and polishing.

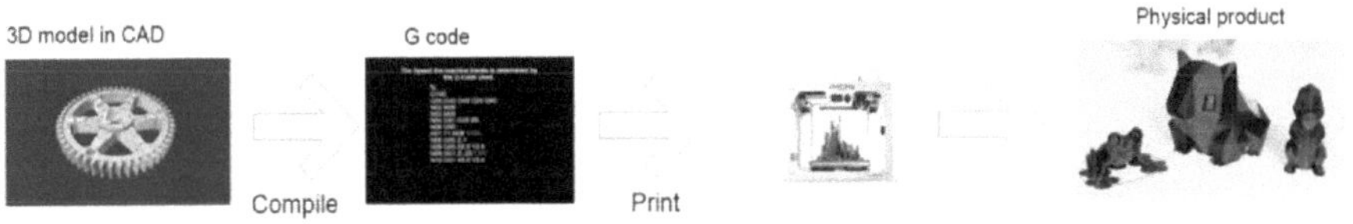

FIGURE 6.1 Processing steps in 3D printing methodology.

6.3 3D PRINTING TECHNIQUES

This sub-section briefly discusses four popular 3D printing processes used for the production of different parts.

6.3.1 STEREOLITHOGRAPHY

STL is a photo polymerization process that makes use of a laser beam to print different section of solid model, layer by layer over photo curable resin surface [1]; the process is considered as cheaper among other techniques. Figure 6.2 shows schematic representation of STL printing process. Some of the notable advantage of STL technique low-cost 3D printing techniques are better dimensional accuracy, user-friendly, high printing speed, improved surface finish, production of watertight parts and isotropic material behavior [2]. The main disadvantage of this process is comparatively slow printing speed.

6.3.2 FUSED DEPOSITION MODELING (FDM)

It is also called fused filament fabrication (FFF) and a popular 3D printing technique developed by Stratasys Inc., employs the principle of extrusion [3]. It mainly used to produce parts of thermoplastic filaments; the filaments are usually cylindrical of 1.75 mm diameter. The filament is heated at one end up to a semi-liquid state that comes out through the nozzle and ejected on the bed plate in layer-by-layer manner. Polylactic Acid (PLA), nylon, PETG, ABS is different materials used for production of parts. The merits of the process are easy material change, low maintenance cost, high production rate, fabrication of complex shape part, and low temperature operation.

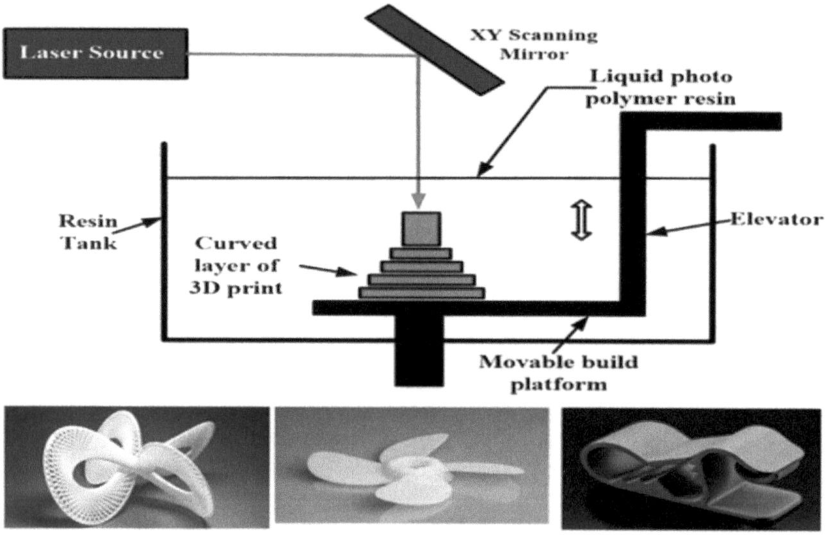

FIGURE 6.2 Schematic diagram of the SLA process and parts produced by it [10].

FDM also has few limitations such as comparatively poor surface finish, slow printing speed, restricted dimensions, etc. [4, 5]. Figure 6.3 shows schematic representation of FDM technique and Figure 6.4 shows different steps involved in printing of 3D print components.

UltimakerCura software (version 3.1.2) is commonly used for preparation of 3D print components. It isa popular open-source slicing software. The part loaded in Cura then desired parameters are set. Then Cura works by slicing the part into specified thickness of the layers and generate G-code to print the component. The generated G-code is sent to the 3D printer for printing/ production of 3D components. The different steps involved are as follows:

- Load STL file of the component to be produced;
- Set the selected process parameters for printing;
- Tool path is generated through layer slicing mechanism;
- Generation of G-code;
- Printing of 3D parts.

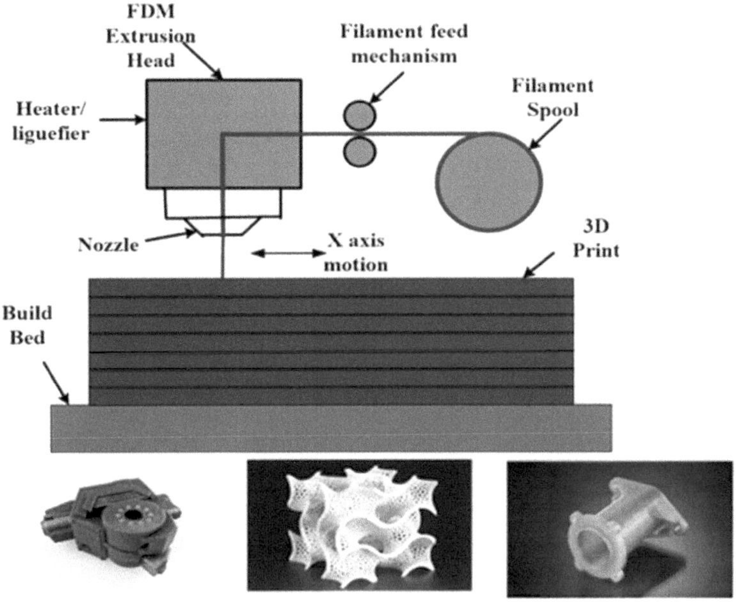

FIGURE 6.3 Schematic diagram of the FDM process and parts produced by it [11].

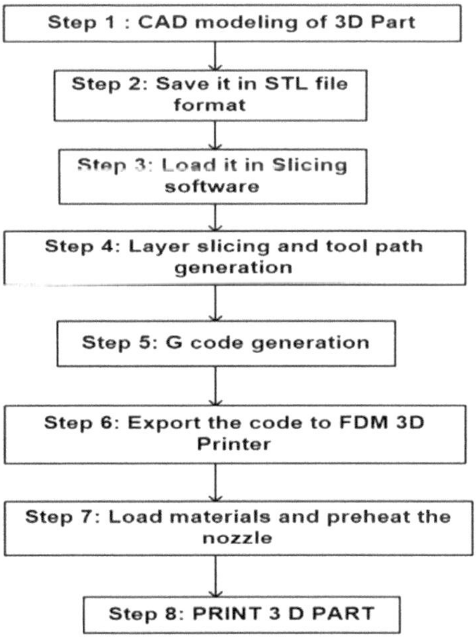

FIGURE 6.4 Processing steps of 3D print components.

Figure 6.5 shows the pictorial view of STL file loading in UltimakerCura software and generation and printing of 3D part.

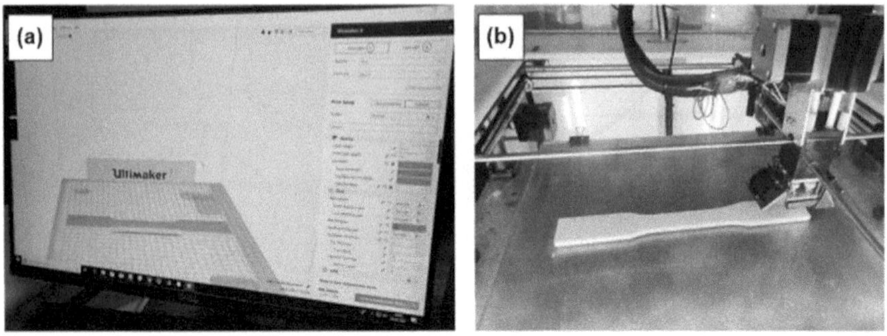

FIGURE 6.5 Production of 3D part (a) STL file loaded, and parameter set in UltimakerCura software; and (b) printed 3D part.

6.3.3 SELECTIVE LASER SINTERING (SLS)

In recent days selective laser sintering (SLS) is another popular technique with a capability to form solid parts of arbitrary geometry without part-specific tooling [7]. In this process, 3D parts are produced in a layer-by-layer manner through the consolidation of powder using a focused laser beam that selectively scans the surface. Laser sintering used to produce parts made of polymeric materials like ABS, nylon, glass-filled nylon, polycarbonate plastics and metals like various steels, copper, and super alloys.

There are two methods viz., direct laser sintering and indirect laser sintering; in which the latter process metal powders are sintered by either fusing with a binder or coating with a polymer. The process involves are consolidation of powder particles by the melting and binding capacity of polymers having low melting point, which are eventually burnt off. The main advantage of direct method laser sintering (DMLS) is to produce metal complex-shaped components in a single step; however, the process requires careful control on temperature variation and densification ratio during the process. This affects distortion part and generate internal stresses; the risk of balling and dross formation in the melt pool reduce surface roughness of the parts produced [6]. The use of SLS is limited for application of parts having low dimensional accuracy those conventional production methods [7]. Figure 6.6 shows a schematic diagram of SLS process.

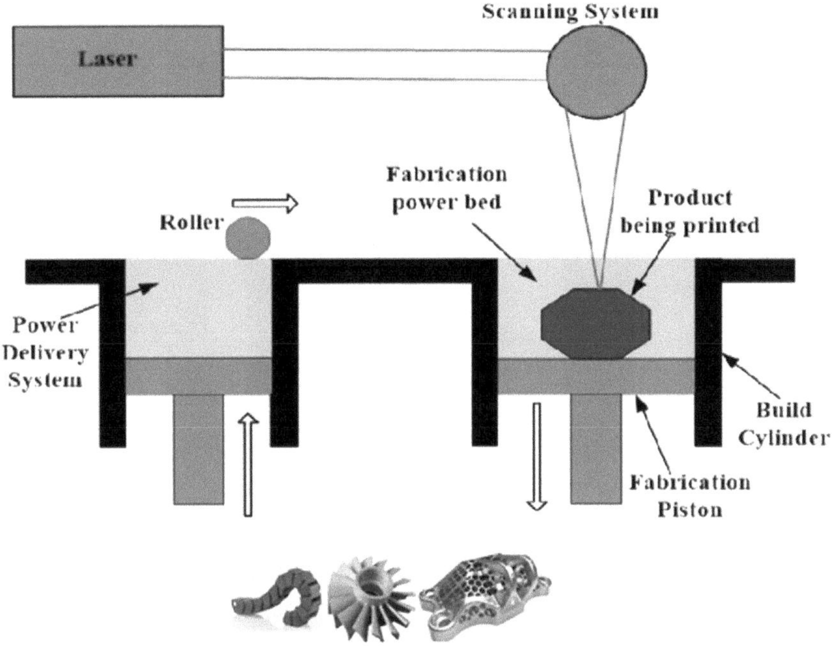

FIGURE 6.6 Schematic diagram of SLS process and parts produced by it [12].

6.3.4 *SELECTIVE LASER MELTING (SLM) TECHNIQUE*

SLM is mainly used for building metallic parts; the process is popular due to its capability fabricating metallic parts with complexity directly from a CAD model. The process involves layer wise deposition of metal using high-power laser beam scans over a metallic powder bed which selectively melts the powder. The molten powder solidifies and cools rapidly forming a layer of the product to be produced. Metal powder is added on completion of each layer and laser scanning is carried out continuously to produce the subsequent layers until the final product is obtained. The non-melted powder in the bed acts both as a support and a medium for heat conduction. The extreme temperature during the process generates residual stresses which require supports so that the part does not move in the powder bed or collapse significantly. This technique is more suitable for production of complex geometrical features [8]. SLM process produces parts with wide range of metals and metal alloys such as aluminum, stainless steel, titanium, cobalt chrome, and Inconel [9]. Figure 6.7 represents schematic diagram of SLM process.

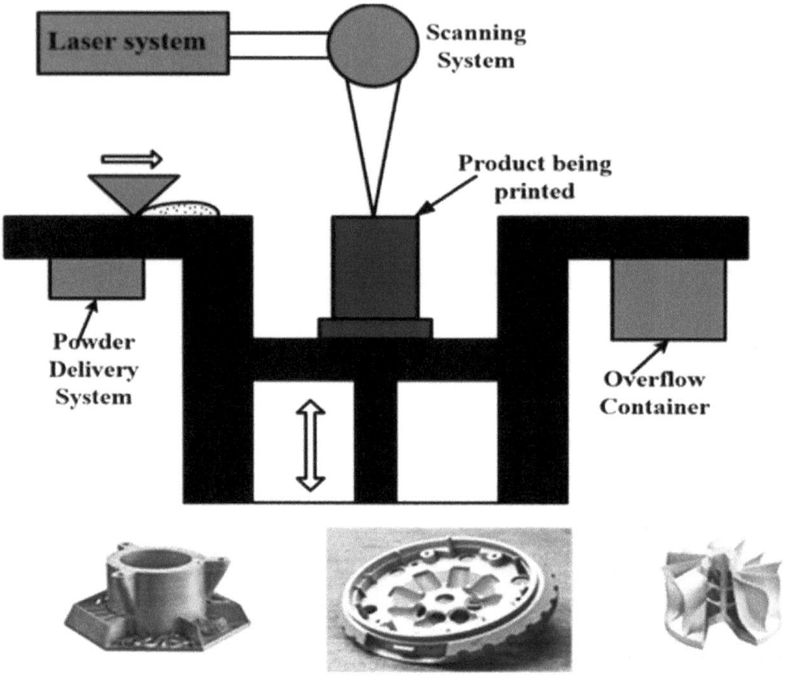

FIGURE 6.7 Schematic diagram of SLM process [13].

Table 6.1 presents the details of different 3D printing techniques and its comparison including their characteristics.

6.4 LITERATURE REVIEW ON 3D PRINTING

Several researchers have carried out much notable research in the field of additive manufacturing by opting various 3D printing techniques as their method of research. Some of them with various 3D printing techniques are mentioned in subsections.

6.4.1 LITERATURE ON SELECTIVE LASER MELTING

Rafi et al. fabricated Ti-6Al-4V and 15–5 PH steel samples using SLM and also analyzed the tensile, fatigue, and fracture properties of the samples and compared them with respect to build orientation. The fabricated two specimen; one is Ti64 with chemical composition (in weight percent): 5.5

TABLE 6.1 Details of Different 3D Printing Techniques

SL. No.	3D Printing Method	Basic Principle	Materials Used; Application	Merits/Demerits	Remarks
1.	Stereo lithography	The laser beam is concentrated on the object to be produced to perform curing and solidification in a layer-by-layer manner.	Plastic and similar materials; jewelry.	• Good surface finish. • High accuracy. • Slow printing speed.	Can be used for low volume production.
2.	Selective laser sintering	Melting of nylon-based powder into solid plastic.	Nylon based powder; high strength prototype.	• High strength. • No support required by the object. • Poor surface finish and dimensional accuracy.	Can be used to print prototype with high strength.
3.	Polyjet	The sprayed molten material is immediately cured by UV laser.	Photopolymers and thermoplastics. Complex shaped prototype with high accuracy.	• Prototype having multiple materials can be printed. • High accuracy and surface finish.	Useful to make prototype with different color and different material.
4.	Digital light processing.	The working is same as that of SLA with the difference that the curing action is done by a digital light projector.	Plastic, etc.; Used for low volume production of plastic prototype.	• Faster build speeds.	Useful for low volume production of plastic prototype.
5.	MultiJet fusion	The working is similar to that of SLS except the curing is done by an inkjet array followed by a heating element in MJF.	Nylon based powder; used to melt nylon powder to hard plastic objects.	• Faster build time. • Low production cost.	Used to make high strength prototype with good surface quality.
6.	Fused deposition modeling	Involves extrusion of plastic filament in a layer-by-layer manner.	Plastic, etc.; used to make plastic rapid prototypes with less strength.	• Faster production rate. • Poor surface finish and strength. • Cost effective.	Can be used for making plastic prototype with less strength.

TABLE 6.1 *(Continued)*

SL. No.	3D Printing Method	Basic Principle	Materials Used; Application	Merits/Demerits	Remarks
7.	Metal 3D printing	It works on reducing sub-parts and to form the product in single piece.	Metal; useful in making complicated metal parts in single piece in medical application.	• High strength product. • Durable product.	Prototypes for medical application is produced using metal 3D printing.
8.	Electron beam melting	It implies electron beam regulated by magnetic coil to print the object.	Metal; used to make metallic component with high melting point.	• High strength metallic prototype is produced. • High surface quality and accuracy.	Metallic prototype with high precision can be printed.

Al, 3.5V, 0.04N, 0.08O, and balance Ti) and the other one is PH1 (chemical composition (in weight percent): 16.4 Cr, 4 Ni, 3.7 Cu, 1 Mn, 1.3 Si, 0.045 C, 0.17 N, and balance Fe) steel powders supplied by EOS. Ti64 is widely used in biomedical implants and aerospace industries due to its high specific strength, stiffness, biocompatibility, and good corrosion resistance. The alloy PH1 stainless steel is commonly used in aircraft components and high-pressure valves, shafts, fasteners, and gears, etc. They studied microstructural images of both the samples with high-magnification scanning electron microscopy–secondary electron (SEM-SE) and found that the samples built horizontally have better tensile properties than that of those built vertically. The tensile and fatigue properties of SLM built samples are found to be comparable to their respective wrought material [8]. Rong-Ji et al. determined the best processing parameters for SLS by minimizing the shrinkage; they implemented the theory and algorithms of the neural network in accordance with their nonlinear and multitudinous processing parameters in order to study the SLS process parameters. They optimized process parameters modeling with artificial neural network (ANN) and optimizing by genetic algorithm approach.

Layer thickness, hatch spacing, laser power, scanning speed, work surrounding temperature, interval time and scanning mode are considered as different process parameters [15]. Dingal et al. carried out experimental investigation of SLS of metallic powder (Trade name: Atomet 86) considering seven process parameters namely, laser peak power density, laser pulse on time, laser scan speed, stepping distance, interval spot ratio, size range of iron powder particles and powder layer thickness as per Taguchi L8 design approach. They found that all these parameters have considerable effect on output characteristics like density, porosity, and hardness [16]. Kuo et al. investigated influence of various process parameters including laser power, exposure duration and point distance on the material density, porosity of the printed body and also dimensional accuracy in 3D printing of Ti-6Al-4V powders in Titanium plates using selective laser melting (SLM). They used Taguchi L9 and ANOVA techniques to study the variance in output characteristics with respect to different weightages of input process parameters [17]. Cai et al. carried out printing of polyamide 12 (PA12) parts using SLS and MJF techniques and did a comparison between the mechanical performance and printing characteristics of printed objects. They concluded that part produced using MJF technique has higher mechanical strength and the part produced using SLS technique has a better surface finish on the top, but the MJF printed product has better surface finish on the front and side surfaces [18]. Chatterjee et al. applied statistical design of experiments to the process to Selective Laser Sintering (SLS) on low carbon steel powder. They investigated the effects of layer thickness and hatching distance on the density, hardness, and porosity of the sintered products by implementing regression analysis. They observed that increasing layer thickness and hatching distance results in an increase in porosity which reduces the hardness and density [19].

6.4.2 LITERATURE ON STEREOLITHOGRAPHY

Zhou et al. inspected a specially designed 20 dimensional, geometrical, and surface roughness features by developing an orthogonal array of experiments with least number of experimental runs and desired process parameter settings. They used RSM and ANOVA to study and optimize the SLA RP process and concluded optimal setups of SLA manufacturing for both individual and general part with various features [20]. Munprom et al. investigated the effect of seven important process parameters viz., exposure resolution, layer height, exposure time, lift, up speed, post cure time and post cure temperature on the object qualities namely flexural stress, hardness, and part dimension. They analyzed the significance of the process parameters and also optimized them by using Taguchi method with the L_8 [27] orthogonal array and reduced the number of experiments and hence reduces the cost and time [21]. Cho et al. focused on improving the build accuracy by optimally designing the process parameters. The process is modeled by using a multi-layer perception neural network and the developed model is optimized using genetic algorithm to find the optimal process parameters in order to obtain minimum build error. The genetic algorithm gives near optimal values of process parameters in order to obtain a near perfect build accuracy [22]. Cosmi et al. carried out tensile tests for SLA 3D printed specimen using a basic in house designed tester. The test specimens were printed at different orientation on a Formlabs™ Form 2™ using clear V4™ resin. They calculated ultimate strength, Young's modulus, strain at maximum stress and strain at breaking point and found no relationship between printing angle and mechanical characteristics, hence the material printed can be considered as isotropic. When the mechanical properties obtained are compared with that of stated by Formslab™ and they found the obtained maximum stress to be slightly less while Young's modulus was found to be almost same [23].

6.4.3 LITERATURE ON FUSED DEPOSITION MODELING

Mohamed et al. [24] reviewed various research that were done in order to determine and also optimize different process parameters in FDM. They found that the RSM is the most efficient optimization method as the error reported is minimal. They also concluded some important process parameters like air gap, layer thickness, raster angle, raster width, build orientation, etc., are critical in FDM process. Yadav et al. [25] studied the variation of tensile strength of acrylonitrile butadiene styrene (ABS), PETG and multi-material test sample with respect to various process parameters like material density, infill density and extrusion temperature. The fabrication of multi-material was accomplished by merging 50% of ABS and 50% of PETG in a layer-by-layer manner using FDM 3D printing. The infill density and extrusion temperature are optimized for increasing the tensile strength of FDM fabricated units. The claimed an increase in tensile strength up to 4.54% using GA-ANN. Gebisa et al. [26] investigated the effect of process parameters on the tensile properties of the component produced by the FDM technique. They carried out the proposed study on a high performance ULTEM 9085 polymeric material by following full factorial design of experiment in order to investigate the effect of process parameters on the tensile properties of the material. The considered five parameters namely air gap, raster width, raster angle, contour number and contour width and analyzed their effect on tensile properties and concluded that only raster angle has significant influence on the tensile property of the material.

Moradi et al. [27] investigated experimentally the effect of infill pattern on mechanical properties of parts made of polylactic acid (PLA) using FDM. Six type of infill pattern is considered and the variations of mechanical properties such as ductility, stress, young's modulus are studied with respect to each of the pattern and also optimized them. The concluded that the pattern has direct effect on the above-mentioned mechanical properties and the triangular pattern is found to be the most suitable for better mechanical properties.

Rankouhi et al. [28] investigated the influence of two major input parameters namely layer thickness and orientation on the mechanical properties of ABS sample manufactured using FDM. They also carried out a study on the typical support frame of the tensile specimen in order to generalize the outcomes of the experimental work performed. Fractography was employed using a scanning digital microscope in order to analyze the effect of layer thickness on failure. The investigation found that the layer thickness and raster orientation have significant influence on the mechanical properties of the tensile specimen. Camargo et al. [29] investigated the effect of infill and layer thickness on various mechanical properties such as tensile strength, flexural strength, and impact energy of PLA-graphene tensile specimen made using FDM. The design of experiment chosen was central composite design (CCD). The obtained 3D printed parts were found to be exhibiting anisotropic behavior due to the layered production process. They opted for flat orientation and honeycomb infill pattern and concluded that the tensile strength and flexural strength increases while the impact energy decreases with infill.

Dou et al. [30] investigated the influence of certain process parameters such as layer height, extrusion width, printing temperature and printing speed on the mechanical properties of 3D printed tensile specimen made of continuous fiber reinforced PLA composites. The study found that the tensile properties vary inversely with layer height, extrusion width and printing speed and they vary directly with printing temperature [30]. Wang et al. printed PEEK specimen using FDM 3D printing technique and implemented FEA to simulate the melting conditions and fluidity of PEEK in a flow channel with aim focused on better surface quality and improved mechanical properties. The simulated model is further analyzed in order to optimize the printing parameters such as printing temperature, printing speed and layer thickness for future references. They investigation found a set of printing parameters 440°C printing temperature, 20 mm/sec printing speed and 0.1 mm layer thickness for the mentioned objectives [31]. Yao Li et al. applied FDM 3D printing method to print PEEK material and investigated the influence of multi-factor coupling such as printing temperature, printing direction, printing paths and layer thickness on tensile strength, bending strength, crystallinity, and grain size of the printed product. The tensile properties of 3D printed PEEK is found better than conventional PEEK material at optimum input parameters 0.1 mm layer thickness, 180 horizontal printing path at 525°C and also improved bending properties are found in 0.3 mm layer thickness, printing path 45° at a printing temperature of 525°C. It is also claimed that the highest crystallinity exceeds 30% which is similar to that of parts produced with injection molding [32].

Wenzheng et al. investigated the influence of layer thickness and raster angle on the mechanical properties of PEEK sample made using FDM technique. The tensile, compressive, and bending strengths were found maximum for layer thickness of 0.3 mm and raster angle of 0°. They also compared the mechanical properties of 3D printed PEEK sample with ABS sample and claimed 108% improved tensile strength, 114% compressive strength and 115% bending strength while elastic properties seemed quite similar [33]. Dhinesh et al. employed combination of PLA and ABS filaments in various compositions in order to improve the tensile and flexural strength of the printed specimen. They printed 30 number of specimens by blending PLA and ABS in various compositions and compared the mechanical properties of the PLA and ABS blend specimen with the pure PLA and ABS material and the comparison shows significant improvement in mechanical strength in favor of the PLA and ABS blend specimen [14]. Fernandes et al. studied the influence of various 3D printing input parameters such as infill density, extrusion temperature, raster angle and layer thickness on mechanical properties of PLA tensile specimen. They concluded the best mechanical properties for input parameter set of 60% infill density, 220°C extrusion temperature, 0°/90° raster angle and 0.1 mm layer thickness. They also minimized the water absorption by the specimen corresponds to 200°C extrusion temperature and 0.2 mm layer thickness [35].

6.5 DISCUSSION AND FUTURE DIRECTIONS

From the review of literature presented, it is evident that the additive manufacturing or 3D printing is found a great potential for development of the manufacturing process. 3D printing eliminates the loss of material completely and it also gives very high degree of dimensional accuracy. It is also capable of producing parts with very complicated shape with complex inner structure which is very difficult to obtain with conventional manufacturing process. Further the process eliminates post processing operations like finishing, and heat treatment. The following are observations found based literature study:

1. From the review of literature, it is found that PLA and ABS are mostly used as work materials. SLA, SLM, SLS, and FDM are most commonly used different process for printing 3D components using polylactic Acid (PLA), nylon, PETG, ABS, polyether ether ketone (PEEK), and metallic powders of low-carbon steel, stainless steel, copper, and super alloys (Ti-6Al-4V); FDM is applied for large verity of materials.
2. PEEK is a high-performance polymer material which has improved mechanical and thermal properties, and it has extensive application in biomedical, aerospace, and automobiles due to its superior mechanical and thermal properties. 3D printing investigation using

PEEK requires attention and very little research work; investigation on mechanical properties of components produced and parametric optimization to be explored.

3. Infill density, layer height, infill pattern, orientation is some of the process parameters employed in FDM process. The investigation considering other possible process parameters and its effect on mechanical strength of the components is not attempted by many of the researchers is not attempted satisfactorily so far. The application of 3D printed products is found in various important fields such as aeronautics, automotive, and biomedical where the strength of the component plays a vital role.
4. Most of the works are aimed at microstructural study of the printed specimen, very less work has been carried out related to the mechanical strength of the 3D printed object by FDM. The investigation on the effect of process parameters on the mechanical properties of the printed specimen employing FDM approach is not performed extensively by many researchers.
5. The area of research reported in literature mainly focuses on improvement in different 3D printing process, extending the process to verity of materials. In order to improve the product quality and minimize the production cost an experimental investigation, process modeling and optimization of process parameters need to be investigated.
6. For modeling and optimization various statistical and soft computing techniques such as RSM, ANN, fuzzy logic (FL), particle swarm optimization (PSO), genetic algorithm (GA), Jaya algorithm and other nature inspired techniques are reported but these are not attempted by the researchers for 3D printing process. The efficiency of these approaches could handle nonlinear, complex engineering problems with acceptable accuracy, which needs to be investigated.
7. 'Sustainability' is a concept of ability to sustain the manufacturing process for a longer period of time considering environmental, economic, and social characteristics. It is essential to identify different indicators/process responses that focus the three essential characteristics of sustainability.

6.6 CONCLUSION

3D printing (additive manufacturing or rapid prototyping) is an advanced manufacturing technology based on layer-by-layer manufacturing to produce complex part without any loss of material. Selective laser sintering (SLS) and FDM are two popular 3D printing approaches capable of manufacturing parts from a variety of materials. This work reviews published literature in the area of 3D printing employing different techniques *viz*., SLA, SLM, SLS, and FDM. PLA, Nylon, ABS, PEEK, and various metallic powders are used as materials for 3D print components used for various applications in aeronautics, automotive, and biomedical industries. Literature study shows experimental investigation of the process for micro-structural study and mechanical properties evaluation. The research area in development of predictive modeling and optimization of process parameters for production of quality components economically need more focus. Application of soft computing approaches in modeling and optimization of the process and development of existing suitable for printing new materials are need of the hour. 'Sustainability' is another important area to develop the 3D printing process considering environmental, economic, and social characteristics; need to be explored.

KEYWORDS

- **3D printing**
- **acrylonitrile butadiene styrene**
- **fused deposition modeling**
- **polyethylene terephthalate glycol**
- **polylactic acid**
- **selective laser melting**
- **stereo lithography**

REFERENCES

1. Zhou, J. G., Herscovici, D., & Chen, C. C., (2000). Parametric process optimization to improve the accuracy of rapid prototyped stereolithography parts. *International Journal of Machine Tools and Manufacture*, *40*(3), 363–379.
2. Francesca, C., & Alberto, D. M., (2020). A mechanical characterization of SLA 3D-printed specimens for low-budget applications. *Materials Today: Proceedings*. https://doi.org/10.1016/j.matpr.2020.04.602.

3. Sandeep, D., Rajan, N., & Deepak, C., (2019). *International Journal on Interactive Design and Manufacturing (IJIDeM), 13*, 1197–1214. https://doi.org/10.1007/s12008-019-00536-z.
4. Ismail, D., & Rukiye, E., (2014). Experimental investigation of FDM process for improvement of mechanical properties and production cost. *Rapid Prototyping Journal, 20*(3), 228–235.
5. Yadav, D., Chhabra, D., Kumar, G. R., Akash, A., & Ashish, P., (2019). Optimization of FDM 3D printing process parameters for multi-material using artificial neural network. *Materials Today: Proceedings*. https://doi.org/10.1016/j.matpr.2019.11.225.
6. Calignano, F., Manfredi, D., Ambrosio, E. P., Iuliano, L., & Fino, P., (2013). Influence of process parameters on surface roughness of aluminum parts produced by DMLS. *The International Journal of Advanced Manufacturing Technology, 67*, 2743–2751.
7. Rong-Ji, W., Xin-Hua, L., Qing-Ding, W., & Wang, L., (2008, 2009). Optimizing process parameters for selective laser sintering based on neural network and genetic algorithm. *Int. J. Adv. Manuf. Technol., 42*, 1035–1042. doi: 10.1007/s00170-008-1669-0.
8. Khalid, R. H., Thomas, L. S., & Brent, E. S., (2013). A comparison of the tensile, fatigue, and fracture behavior of Ti-6Al-4V and 15-5 PH stainless steel parts made by selective laser melting. *Int. J. Adv. Manuf. Technol., 69*, 1299–1309. doi: 10.1007/s00170-013-5106-7.
9. Mingming, M., Zemin, W., Ming, G., & Xiaoyan, Z., (2015). Layer thickness dependence of performance in high-power selective laser melting of 1Cr18Ni9Ti stainless steel. *Journal of Materials Processing Technology, 215*, 142–150.
10. Raju, B. S., Chandrashekar, U., Drakshayani, D. N., & Chockalingam, K., (2010). Determining the influence of layer thickness for rapid prototyping with stereolithography (SLA) process. *International Journal of Engineering Science and Technology, 2*(7), 3199–3205.
11. Haidiezul, A. H. M., Aimanand, A. F., & Bakar, B., (2017). Surface finish effects using coating method on 3D printing (FDM) parts. *IOP Conference Series: Materials Science and Engineering, 318*, 012065. doi:10.1088/1757-899X/318/1/012065.
12. An-Nan, C., Jia-Min, W., Kai, L., Jing-Yan, C., Huan, X., Peng, C., Chen-Hui, L., & Yu-Sheng, S., (2018). High-performance ceramic parts with complex shape prepared by selective laser sintering: A review. *Advances in Applied Ceramics, 117*(2), 100–117. doi: 10.1080/17436753.2017.1379586.
13. Qiang, C., Gildas, G., Charles-André, G., & Michel, B., (2021). Three-dimensional finite element thermomechanical modeling of additive manufacturing by selective laser melting for ceramic materials. *Additive Manufacturing*. doi: 10.1016/j.addma.2017.02.005.
14. Fernandes, J., Deus, A. M., Reis, L., Vaz, M. F., & Leite, M., (2018). Study of the influence of 3D printing parameters on the mechanical properties of PLA. *Proceedings of the 3rd International Conference on Progress in Additive Manufacturing (Pro-AM 2018)*, 547-552. doi: 10.25341/D4988C.
15. Rong-Ji, W., Xin-Hua, L., Quing-Ding, W., & Wang, L., (2009). Optimizing process parameters for selective laser sintering based on neural network and genetic algorithm. *International Journal of Advanced Manufacturing Technology, 42*, 1035–1042.
16. Dingal, S., Pradhan, T. R., Sarin, S. J. K., Roy, C. A., & Roy, S. K., (2008). The application of Taguchi's method in the experimental of the laser sintering process. *International Journal of Advanced Manufacturing Technology, 38*, 904–991.

17. Chuliang, K., Chihhao, S., & Anchun, C., (2017). Parametric optimization of density and dimensions in three-dimensional printing of Ti-6Al-4V powders on titanium plates using selective laser melting. *International Journal of Precision Engineering and Manufacturing, 18*(11), 1609–1618.
18. Chao, C., Wei, S. T., Jiayao, C., Wei, Z., Xingjian, L., Tong, L., Lihua, Z., & Kun, Z., (2021). Comparative study on 3D printing of polyamide 12 by selective laser sintering and multi-jet fusion. *Journal of Materials Processing Tech., 288*, 116882.
19. Chatterjee, A. N., Sanjay, K. P., Saha, P., Mishra, P. K., Roy, A., & Choudhury, A., (2003). An experimental design approach to selective laser sintering of low carbon steel. *Journal of Materials Processing Technology, 136*, 151–157.
20. Jack, G. Z., Daniel, H., & Calvin, C. C., (2000). Parametric process optimization to improve the accuracy of rapid prototyped stereolithography parts. *International Journal of Machine Tools & Manufacture, 40*, 363–379.
21. Munprom, R., & Limtasiri, S., (2019). Optimization of stereolithographic 3D printing parameters using Taguchi method for improvement in mechanical properties. *Materials Today: Proceedings, 17*, 1768–1773.
22. Cho, H. S., Park, W. S., Choi, B. W., & Leu, M. C., (2000). Determining optimal process parameters for stereolithography process via genetic algorithm. *Journal of Manufacturing Systems, 19*(1).
23. Francesca, C., & Alberto, D. M., (2020). A mechanical characterization of SLA 3D-printed specimens for low-budget applications. *Materials Today: Proceedings.*
24. Omar, A. M., Syed, H. M., & Jahar, L. B., (2015). Optimization of fused deposition modeling process parameters: A review of current research and future prospects. *Adv. Manuf., 3*, 42–53. Doi: 10.1007/s40436-014-0097-7.
25. Aboma, W. G., & Hirpa, G. L., (2019). Influence of 3D printing FDM process parameters on tensile property of ULTEM 9085. *Procedia Manufacturing, 30*, 331–338.
26. Mahmoud, M., Ahmad, A., & Alireza, H., (2021). Experimental investigation on mechanical characterization of 3D printed PLA produced by fused deposition modeling (FDM). *Mater. Res. Express, 8*, 035304. https://doi.org/10.1088/2053-1591/abe8f3.
27. Behzad, R., Sina, J., Fereidoon, D., & Todd, L., (2016). Failure analysis and mechanical characterization of 3D Printed ABS With respect to layer thickness and orientation. *J. Fail. Anal. And Preven., 16*, 467–481. doi 10.1007/s11668-016-0113-2.
28. Jose, C. C., Lesson, A., Machado, R., Erica, C. A., & Erickson, F. M. S. S., (2019). Mechanical properties of PLA-graphene filament for FDM 3D printing. *The International Journal of Advanced Manufacturing Technology, 103*, 2423–2443. https://doi.org/10.1007/s00170-019-03532-5.
29. Hao, D., Yunyong, C., Wenguang, Y., Dinghua, Z., Junjie, L., Zhoujun, M., & Stephan, R., (2020). Effect of process parameters on tensile mechanical properties of 3D printing continuous carbon fiber-reinforced PLA composites. *Materials, 13*, 3850. doi: 10.3390/ma13173850.
30. Peng, W., Bin, Z., Hongchuan, X., Shouling, D., & Chuanzhen, H., (2019). Effects of printing parameters of fused deposition modeling on mechanical properties, surface quality and microstructure of PEEK. *Journal of Materials Processing Technology, 271*, 62–74. https://doi.org/10.1016/j.jmatprotec.2019.03.016.
31. Yao, L., & Yan, L., (2020). Tensile and bending strength improvements in PEEK parts using fused deposition modeling 3D printing considering multi-factor coupling. *Polymers, 12*, 2497. doi: 10.3390/polym12112497.

32. Wenzheng, W., Peng, G., Guiwei, L., Di, Z., Haibo, Z., & Ji, Z., (2015). Influence of layer thickness and raster angle on the mechanical properties of 3D-Printed PEEK and a comparative mechanical study between PEEK and ABS. *Materials, 8*, 5834–5846. doi: 10.3390/ma8095271.
33. Dhinesh, S. K., Arun, P. S., Senthil, K. K., & Megalingam, A., (2021). Study on flexural and tensile behavior of PLA, ABS and PLA-ABS materials. *Materials Today: Proceedings, 45*, 1175–1180.

CHAPTER 7

APPLICATIONS OF MEDICAL IMAGE PROCESSING AND COMPUTER VISION IN HEALTHCARE

S. SUDHAGAR[1], S. SURESH[1], P. M. GOPAL[2], and S. SEENIVASAN[3]

[1]*Department of Mechanical Engineering, Erode Sengunthar Engineering College, Erode, Tamilnadu, India*

[2]*Center for Materials Science, Mechanical Engineering, Karpagam Academy of Higher Education, Coimbatore, Tamil Nadu, India*

[3]*Department of Mechanical Engineering, Dhaanish Ahmed Institute of Technology, Coimbatore, Tamil Nadu, India*

ABSTRACT

There is great potential for computer vision to improve medical care worldwide. Applications for computer vision drastically cut down on the time and effort needed to detect medical conditions or the effects of medication on a particular population. New algorithms are constantly being proposed by researchers to process and extract characteristics from medical images. The advancement of image processing techniques has led to a notable growth in its contribution to the healthcare area, providing clinicians with additional diagnostic information. Autonomous robotic surgery may become a reality in the near future, reducing medical errors brought on by weariness. The medical industry is benefiting from the development of AR and VR technologies, which can simulate the human anatomical body and teach students with low resources.

Robotics and Automation in Healthcare: Advanced Applications. R. Thanigaivelan, Sanjay Singh, & Clement Christy Deepak C. (Eds.)

7.1 INTRODUCTION

Medical imaging is one of the fundamental and inevitable processes in medical science. Medical imaging process is a non-invasive process of creating images of internal organs. In medical imaging process, the images of inner organs of the body are clearly visualized and which is utilized for understanding human biological system. The medical imaging helps to improve the diagnosis and treatment of a particular disease or internal anomalies. Several imaging techniques are available for acquiring the internal organ images such as X-rays, gamma rays, sonography, magnetic energy, thermal imaging, scopes, etc. Each imaging technique has its own advantages and limitations. These techniques convert the energies into arithmetic and logic signals, which are then converted into digital images [1, 2].

The analysis of medical images visually by human might have a chance for error because of the interpretation of information is limited to observers' experience, expertise, fatigue, and distraction. Hence, medical images are processed using advanced computer programs which is capable of extracting tiny information accurately. Medical image processing is the process of performing various operations on the digital image using computers. Obtaining medical images in digital format have several advantages such as easy storing and retrieval, instant quality assessment, better communication, lower processing cost, copying, and multiple reproduction. The medical images in digital format are the mathematical functions of measured values from imaging system. In computer Tomography, the coefficient of absorption measured is converted into gray values. These gray values are stored with greatest accuracy, since numerous values of gray shades of grays are possible between black and white. Identifying small details in these gray values are difficult for human eye. The image processing operations involves image storing, removing noises, detecting edges and sizes of objects, etc. [3, 4].

The following section of this chapter discuss various methods adopted in healthcare industries for imaging the internal organ, different steps involved in processing of captured image and various application in which the medical image processing are playing important role.

7.2 STEPS OF IMAGE PROCESSING

The medical images are involved in various digital image processing steps to extract information from image and using that information the diagnostics and treatments are carried out [5]. Figure 7.1 shows the various steps involved in medical image processing.

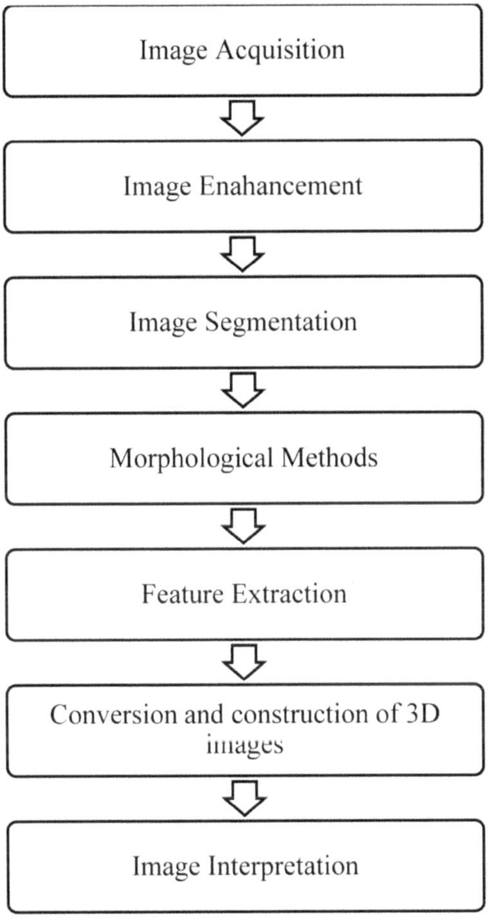

FIGURE 7.1 Steps involved in image processing.

7.2.1 IMAGE ACQUISITION

Image acquisition is the basic and foremost step in medical image processing. It is the process of capturing the image of internal organs for study. There numerous methods are available for capturing the medical images such as X-ray imaging, Ultrasound, Magnetic Resonant Imaging, Computer

Tomography, etc. Each imaging technique has its own features and advantages, which are discussed in further sections.

7.2.2 IMAGE ENHANCEMENT

The image captured through various techniques may contain eternal noises and disturbances and these should be eliminated before interpreting and gathering information from medical image. In image enhancement step, the noises present in the image are removed and increase the visual representation of image.

7.2.3 IMAGE SEGMENTATION

Image segmentation is the initial preprocessing step in all kinds of biomedical image processing. The image segmentation is the process of subdividing the acquired medical image smaller segments or regions. The segmentation of image into smaller regions will be helpful in performing certain task like identification of different organs such as heart, lungs, liver kidney, etc., in the image. The application of image segmentation is not limited to organ detection, but it can be also utilized to detect and label brain structures such as thalamus or the hippocampus. Image segmentation also useful in quantitative measurement of ultrasonic image. The image segmentation of medical image can be classified into two approaches based on features and techniques [6].

The feature-based image segmentation is made by utilizing the image features such as intensity of pixels, magnitude of gradients and measure of texture. Various segmentation techniques utilizing these features are broadly classified into three categories namely edge based technique, region-based technique and classification. The edge based and region-based techniques of image segmentation is achieved by identifying the pixels within-region similarities and between-region differences between features. The classification technique of image segmentation is achieved by assigning labels to individual pixels based on the features of pixels.

7.2.4 THRESHOLDING

Thresholding is an easy and effective way of image segmentation in medical image processing. In Thresholding approach, gray level value of each object or structure or pixel in the image is compared with the one or more intensity threshold values and assign labels to the pixels. If one value of threshold has been used for image segmentation, this technique segments the image into two regions namely foreground and background. The foreground label is assigned to a pixel if gray level value of the respective pixel is higher than the threshold value otherwise the pixel is labeled as background. However, in medical image processing the image need to be segmented into multiple regions. In order to achieve this, multiple threshold values are used.

7.2.5 EDGE-BASED TECHNIQUE

In this, technique the different regions of processed image are defined by its boundaries or edges. The intent of image segmentation is to identify, and extract required structure from the image, but the edge detection algorithm only generates a series of boundary data respective to interest region. These boundary data are to be properly linked with each other to produce well defined boundaries of structure. There are numerous edge detection algorithms are available and most of them are named after their inventors. The most popular edge detection algorithm is Marr-Hildreth (Laplacian-of-Gaussian), Sobel, Prewitt, and Canny operators.

7.2.6 REGION-BASED TECHNIQUE

The region-based technique of image segmentation involves separation of the image into several regions based on certain homogeneity property. Most commonly using homogeneity criteria for image segmentation is average intensity value in region, texture, or color of image. In this technique, each pixel is evaluated using a set of conditions and based on the satisfaction criteria the pixel is assigned to particular region. The region-based image segmentation algorithm broadly classified into three categories namely region growing, region splitting, and split-and-merge.

7.3 MORPHOLOGICAL METHODS

Morphological methods in image processing are a set operation that process the image based on the shape. Morphological operations are used to remove smaller object in the image while preserving the shape of larger object. In morphological operation, a matrix or small size template called structuring element used for processing the image. The structuring element is navigated to all possible position in the image and the image pixels directly linked to the structuring element is analyzed. The structuring element can be selected to any shape and size based on the requirement. Four major operations are involved in morphological method namely dilation, erosion, opening, and closing. The dilation operation adds pixels to the object boundary relative to structure element whereas erosion operation removes pixels from the object boundary. The opening and closing operation are compounding of dilation and erosion and the difference between them is sequence of operation carried. In closing operation, initially dilation is performed on image followed by erosion operations, whereas in opening operation, vice versa is performed.

7.4 FEATURE EXTRACTION

Features are the useful information extracted from the image which is utilized for diagnosis. The features extracted from biomedical image may be based on individual pixels, areas, time, and transformation of data. Features play a number of roles in medical image analysis. The features extracted from biomedical image is helpful for various process in medical field such as disease detection, diagnosis, therapy, and treatment planning. The features extracted from medical image are used to detect the change in size of internal anatomical structure, identifying the type of blood cells, counting number of blood cells and identification of treatment areas. Many kinds of transforms are usefully applied for feature extraction. The Fourier transform, for example, provides information about spatial frequency for the image or some subimage.

7.5 CONVERSION AND CONSTRUCTION OF 3D IMAGES

Many situations in clinical practice requires 3D geometric model of different anatomical structure. A wide range of medical imaging applications necessitate geometric representations of the surface of distinct anatomical components. Surfaces are particularly useful for demonstrating the geometric relationships between distinct structures. Surface representation of anatomical structure requires different geometric objects. One method of surface representation in medical image is by using the point set or vertex. The point set consists of cartesian coordinate (x, y, z) of each point on the surface.

7.6 IMAGE INTERPRETATION

The main purpose of biomedical image processing is to assist and aid the human professional to discover and infer maximum possible useful information from the image. A medical image interpretation system creates an internal structure of the anatomy featured in the image so that it may be analyzed anatomically instead of using it as an array of gray levels. This helps the image interpretation system to apply anatomical information into its image analysis that go beyond traditional image processing approaches. A medical image interpretation system applies information particular to the imaging technology and the anatomy being scanned. This medical image interpretation system could act as assist for solving medical imaging problem such as assessing the volume of tumor tissue response to therapy or identifying the type of tumor.

7.7 MEDICAL IMAGING SYSTEMS

7.7.1 X-RAY IMAGING SYSTEMS

X-rays are electromagnetic rays; thus, they follow the same laws as other electromagnetic radiation. Electromagnetic radiation, like radio waves, visible light, or microwaves, carries energy, also known as radiant energy, through space via waves and photons. Numerous advancements are happened in X-ray imaging systems which makes it more reliable. Unlike traditional X-ray films that employ X-rays to straightly alter the chemical characteristics of the film, current detecting systems alter X-rays into light and then as electrons [7]. Some important techniques that are followed in X-ray imaging systems are discussed in detail. The schematic of X-ray imaging system is shown in Figure 7.2.

7.7.2 IMAGE INTENSIFIERS

X-ray image intensifiers are based on vacuum tubes which converts X-rays as a visible light, resulting in an image and the same is illustrated in Figure 7.3. The X-ray photons that are entering inside are first converted to light photons with the aid of an input phosphor made up of phosphorus substance. The photoelectric effect inside a photocathode is used to transform the generated light into electrons. With the aid of an electron optic system, these matters are then speeded up and pointed on the output phosphor. Then it is converted again as a visible light in the output phosphor that may be seized by an image capturing devices or film material.

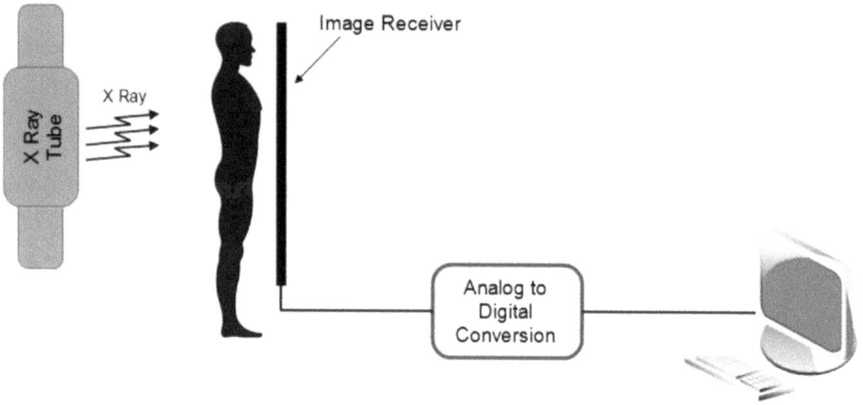

FIGURE 7.2 Schematic representation of X-ray imaging system.

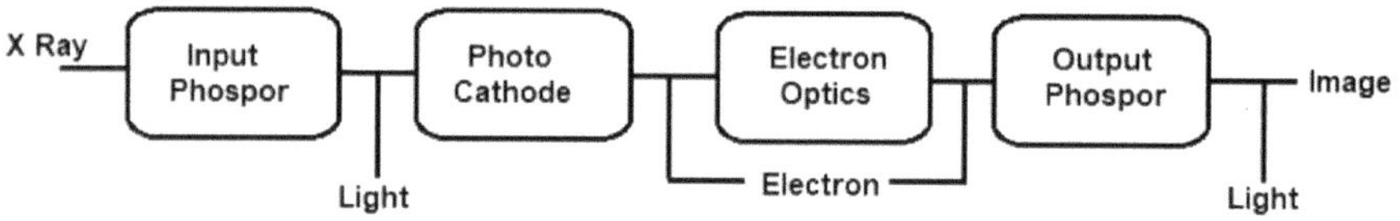

FIGURE 7.3 Schematic principle of image intensifier sensor.

7.7.3 FLAT PANEL DETECTORS

For radiography, angiography, and C-arm CT applications, the flat panel detector (FPD) has recently become the high-tech in X-ray indicator technology. Their key advantages include a straight digital data of the X-ray output with enhanced spatial resolution. They were originally launched in the mid-1990s. Direct and indirect conversion FPDs are the two types of FPDs.

7.7.3.1 INDIRECT CONVERSION FPDS

Indirect FPD uses a layer of cesium iodide (CsI) to change X-rays to light photons like image intensifier system. In addition, the tubular construction of CsI is also similar to the input layer of an image intensifier system. The ensuing detection steps are the most significant difference. Image intensifiers use a process in which light photons are converted to electrons, which are then speeded up to amplify and regulate lighting. FPDs do not require this extra conversion step, as an alternative, an array of photodiodes is straightly linked to the CsI layer and transforms the discharged photons into an electric charge. For each pixel, this electric charge is deposited in capacitors. Additionally, a thin-film transistor (TFT) is included in each pixel, which works as a miniature "switch" for reading the stored charges.

7.7.3.2 DIRECT CONVERSION FPDS

Direct conversion FPDs possess a uniform X-ray sensitive photoconductor layer on the TFT matrix rather than a plain alteration to light photons. A high-voltage bias electrode forms the top layer, which creates an electric field over the photoconductor. When X-rays are engrossed by the photoconductor, electron-hole pairs are liberated. The global electric field separates these couples into negative and positive charges, which are then transferred to the pixel's electrodes. Positive charges are stored in capacitors at the bottom of individual pixel electrodes.

7.8 COMPUTED TOMOGRAPHY (CT)

Computed tomography (CT) is without a doubt one of the most essential medical imaging technologies, providing us with images inside the human body that are both useful and exciting to clinicians. In CT, a thin beam of X-rays is concentrated on a required area and rapidly spun around that area or part, producing signals which are evaluated by the attached processer to develop cross-sectional images—or "slices"—of the particular area or part. These images are slices that contain additional comprehensive data than traditional X-rays. As soon as the processing computed has received the quantity of consecutive slices, it can be stacked together digitally in order to form a 3D image of the particular area or body part which makes it easier to find and trace the fundamental structures in addition to supposed tumors or irregularities.

Contrasting to a standard X-ray, a CT scanner utilizes a motor-powered X-ray source that swirls about the circular entry of a gantry-shaped frame. The person who is going to be diagnosed lies on a bed that travels steadily thru the gantry as an X-ray tube rotates about the bed, shooting thin beams of

X-rays on the person lies. In place of film, these scanners employ digital X-ray sensors which are fixed immediately opposite the X-ray source. Once the X-rays depart the patient, the sensors capture it and directs to the processing unit. Whenever the X-ray source finishes one complete revolution, the CT computer employs complicated mathematical processes to develop a 2D image slice of the scanned part. Depending on CT machine, the visible tissue thickness in individual imaging slice differs, but it generally varies from 1 to 10 mm. As soon as a complete slice has been made, the picture is stored, and then the motor-powered bed is gently moved into the gantry. This process is repeated to create a new image slice. It is repeated till the required quantity of slices are accomplished. The working principle of CT scan is illustrated in Figure 7.4. The skeleton, organs, and tissues in addition to any anomalies the doctor is seeing for, can be displayed separately or heaped together by the computer to produce a 3D depiction of the patient. This approach has numerous positives, comprising the capability to rotate the 3D image in space that facilitates the identification of a problem's precise location [8].

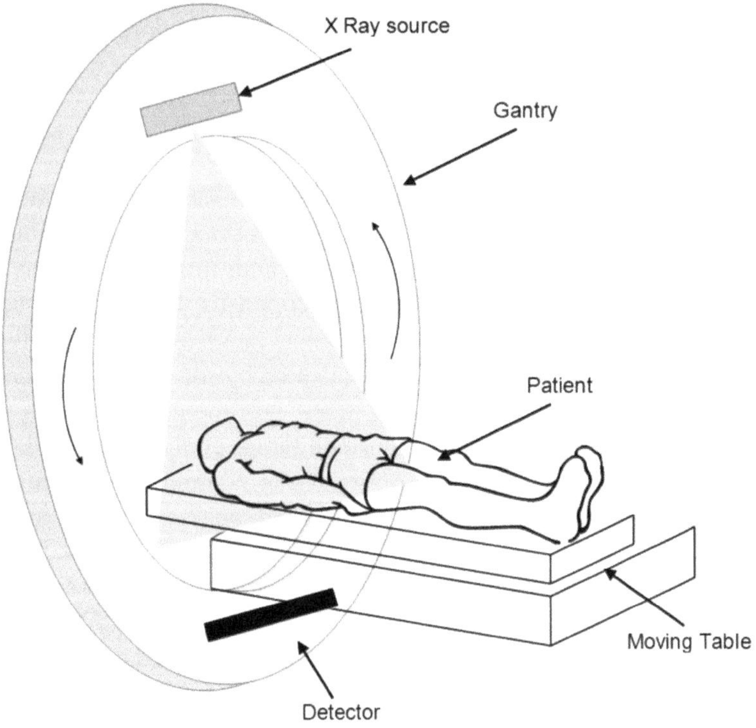

FIGURE 7.4 Schematic of computer tomography.

7.9 NUCLEAR MEDICINE

In nuclear medicine, minor quantity of radioactive chemicals termed radiotracers are passed into the bloodstream, blown in nuclear medicine imaging. A distinct camera is employed to capture the emitted energy in the gamma ray form when the radiotracer injected in the zone being analyzed and the same is used to obtain images of the part. Nuclear medicine imaging is capable of providing data that cannot be given by traditional imaging techniques and it, also possesses the capability of detecting disease even at a premature phase.

Nuclear medicine technique utilizes an exclusive gamma camera and single-photon emission computed tomography (SPECT) imaging method. The radiotracer emitted energy from the body part is seized by the former, i.e., gamma camera and transformed as an image. The gamma cameras utilized in this system is merely a radiation detector and in its own it does not release any radiation. The complete setup is bounded with metal and plastic body, which is combined to a circular, donut-shaped frame, often in box shape. Two gamma camera heads are generally placed on the top and bottom side of the table where the body part or the patient to be positioned. The comprehensive 3D images were created by the spinning action of these camera heads around patient's body in SPECT. The multi-ring detectors positioned inside the system records the radiotracer emitting energy and the detected data is converted into image with the aid of a computer.

The nuclear medicine stands alone from the other imaging models by concentrating on the process happening inside of the body such as metabolic rates and stages of further chemical actions. "Hot spots" are zones with higher concentration which may disclose high radiotracer concentrations besides a large amount of chemical or metabolic action. On the other hand, less concentrated zones, or "cold spots," indicate lower radiotracer strength and chemical action.

Positron emission tomography (PET) is a key nuclear medicine technique that assesses the metabolic activity of cells in bodily tissues, and it is an amalgamation of nuclear medicine and biochemical. PET is most typically employed to visualize biochemical changes in the body of patients with brain or heart diseases, as well as cancer.

7.10 ULTRASOUND

Human hearing can detect acoustic waves with frequencies between 16 Hz and 20 kHz, which are referred to as audible waves or audible sound.

Ultrasound is defined as a frequency greater than 20 kHz. Because of its inexpensiveness and lack of ionizing radiation, medical ultrasound is now often used as a first-line clinical imaging modality. Medical ultrasound frequencies typically range from 2 MHz to 40 MHz. Medical ultrasound has traditionally been used for diagnostic purposes, although additional therapeutic applications are emerging.

An ultrasound transducer is a device that can both generate and detect ultrasonic waves. Mechanical energy is converted to electrical energy and vice versa. High-frequency sound waves are directed into the body when the transducer is pushed against the skin. Because the transducer's sound waves can only just penetrate air, a gel is employed on the skin to reduce the quantity of air among the transducer and the skin. As a result of reflection and scattering of penetrated waves, the body fluids and tissues produces the sound echoes which is recorded by the transducer. Based on the strength and quality of these echoes 1D, 2D, or 3D images are created based on the nature of transducer used [9].

Transducers uses piezoelectric phenomenon to produce and sense the ultrasonic waves. In piezoelectric materials, it defines the transformation of electrical energy to mechanical energy and the other way around. Mechanical pressure (pressure translates to "piezo" (gr.)) is transferred to electric polarization, which causes electric voltage on the one hand. As demonstrated in Figure 7.5, the electric voltage can be quantified using two electrodes. Electrical fields, on the other hand, cause the piezoelectric material to shrink or stretch. By introducing a high-frequency alternating voltage to this contraction and stretching, ultrasonic waves can be generated.

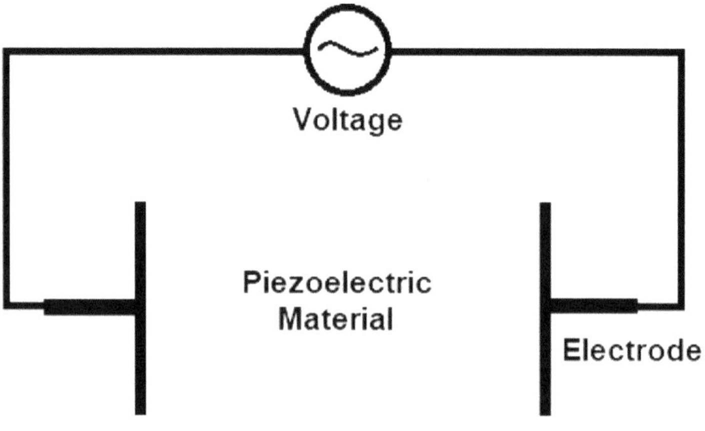

FIGURE 7.5 Piezoelectric effect.

Barium titanate ($BaTiO_3$) and lead zirconium titanate are two common materials utilized in medical ultrasound transducers (PZT). Ultrasound imaging modes include a wide range of options wherein A-mode, B-mode, and M-mode are the most common ones. The A- and M-modes produce 1-D images (signals), while the B-mode can produce 2-D or even 3D images. Doppler can be attained in 1-D, 2-D, and 3D using the latest transducer versions.

7.11 MRI

A magnetic field and radio waves made through computer are utilized in magnetic resonance imaging (MRI) to obtain detailed pictures of patients' organs and tissues. These scanners are particularly well suited for picturizing non-bony body components and soft tissues. It differs from CT by not utilizing X-rays, which releases injurious ionizing radiation. MRI captures the brain, spinal cord, and nerves, in addition to muscles, ligaments, and tendons, significantly much clear than usual scanning technologies such as CT and X-ray; thus, it is often employed to picturize knee and shoulder hitches. It has the ability to identify the variation between white and gray substance in the brain and also be employed to analyze aneurysms and malignancies. Since this technique doesn't involve any X-ray like radiation for diagnosis, MRI is the primary choice when regular imaging is required for treatment.

A strong magnetic field is created with the aid of powerful magnets in MRI, which forces protons in the body to line up with the field. Once a radiofrequency current is pulsed through the body, pushing against the magnetic field's attraction, protons are activated and spin out of balance. If the current is turned off, the protons will realign with the magnetic field wherein the energy produced during this time by protons is captured by the MRI sensors. Based on the surroundings and chemical face of the molecules in the analyzed area, the proton realignment time and quantity of released energy varies which aids the physicians to identify the variation among different types of tissues. In order to obtain the MRI image, the patient is positioned inside a large magnet, and it is mandatory to continue in very static position even without any tiny motion to get clear image without blur. Figure 7.6 shows the layout and component of MRI scan system. To quicken the protons realigning speed with the magnetic field, distinct medicines (generally comprising the Gadolinium element) may be provided intravenously to the patient earlier or during the MRI. The brighter the picture, the quicker the protons realign [10].

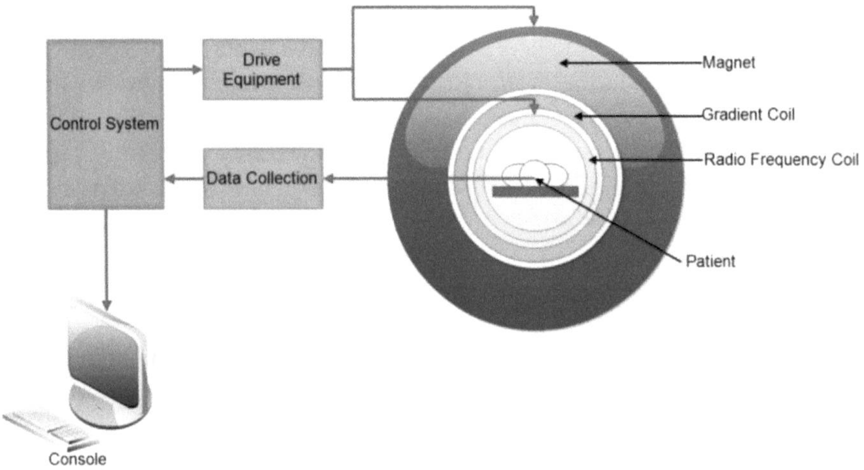

FIGURE 7.6 Schematic of MRI scan.

7.12 OPTICAL COHERENCE TOMOGRAPHY (OCT)

Optical coherence tomography (OCT) is a noninvasive imaging procedure that produces high-resolution cross-sectional images of the retina. To support the premature discovery and analysis of retinal disorders and ailments, the layers within the retina can be divided and retinal thickness can be quantified. For the examination and treatment of most retinal diseases, OCT testing has become the gold standard. OCT measures retinal thickness using light rays. This test does not include any radiation or X-rays, and an OCT scan is not painful or unpleasant. An OCT scan may be performed for a variety of reasons, including monitoring the progression of disease, confirming, or dismissing suspected retinal edema. Coherence Optical Tomography is comparable to ultrasound, wherein it employs light in place of sound to obtain better, sharper resolution [11]. The line diagram illustrated in Figure 7.7 shows the working principle of OCT.

7.13 APPLICATION IN HEALTHCARE

Medical imaging is used significantly in today's healthcare industry to provide precise diagnosis. There have been new medical imaging techniques and procedures implemented. Tumor detection, medical imaging, cancer detection, medical training, fighting COVID-19, continuous health monitoring, timely disease detection, machine-assisted diagnosis, remote patient monitoring, and lean management are all applications where computer vision plays a crucial part [12–14].

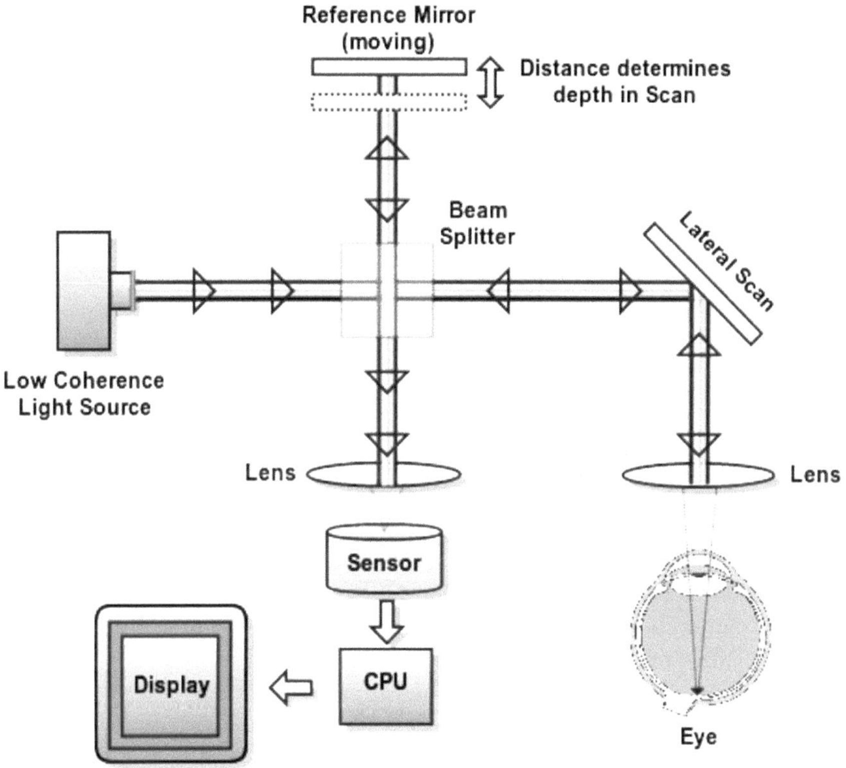

FIGURE 7.7 Schematic of optical coherence tomography.

7.13.1 MEDICAL IMAGE FOR DIAGNOSIS AND SURGERY

Computer vision can help with any medical task that requires a trained eye to identify and classify a health issue. This AI-powered technology employs complex algorithms to aid image processing and reads it in real time, detecting specific indicators of disease. The right application of computer vision in medicine can assist reduce the amount of time spent on unneeded diagnostic procedures while also allowing doctors to make more accurate diagnoses and prescribe more effective therapies [15]. Initially the machine vision for diagnosis and surgery was based on statistical processing of signal, but the advancement in computing technology shifted towards the application of machine learning (ML) algorithm. Analyzing of video is more potential method of diagnosis than image analysis since video contains 25 times more data than an image. A study had explored that analysis of laparoscopic video for diagnosis resulted in 92.8% accuracy in identification. The integration of Artificial Intelligence (AI) with machine vision can made the robot surgeon possible in near future. Smart Tissue Autonomous Robot (STAR) developed by Johns Hopkins University can perform some surgery better than human surgeons in animals. Though the video analysis is a potential method in diagnosis and surgery, it is at premature stage and more exploration is needed in this context.

Computer vision systems have proven to be superior to human specialists at recognizing patterns and detecting diseases without error. By identifying minute changes in tumors, these technologies can assist clinicians in diagnosing malignancy. By scanning medical pictures, such tools can aid in the detection, prevention, and treatment of a number of ailments.

7.13.2 APPLICATION OF COMPUTER VISION IN TUMOR DETECTION

In the medical field, computer vision along with deep learning applications have proven to be quite useful, particularly in the accurate diagnosis of brain tumors. Brain tumors quickly spread to other parts of the brain and spinal cord if untreated, making early detection crucial to the patient's survival. Computer vision techniques are used by medical professionals to quicken and improve the detection process. In the field of medicine, computer vision techniques such as Mask-R Convolutional Neural Networks (Mask R-CNN) can aid in the detection of brain cancers, reducing the risk of human mistake significantly. Figure 7.8 illustrate the application of image processing to detect brain tumor from MRI scan.

7.13.3 APPLICATION OF COMPUTER VISION IN CANCER DETECTION

It is noteworthy that deep-learning computer vision models have obtained doctor-level accuracy at diagnostic tasks like differentiating moles from melanomas. For instance, because the symptoms of skin cancer sometimes resemble those of other skin disorders, it can be challenging to make an early diagnosis. As a solution, scientists have used computer vision technologies to effectively distinguish between cancerous and non-cancerous skin lesions. There are various benefits to employing computer vision and deep learning systems to diagnose breast cancer, according to research. It can help automate the detection process and limit the likelihood of human error by using a large library of photos containing both healthy and malignant tissue. In the near future, healthcare computer vision systems might be used to diagnose various cancers, such as lung and bone cancer, as a result of the quick advancements in technology [17, 18].

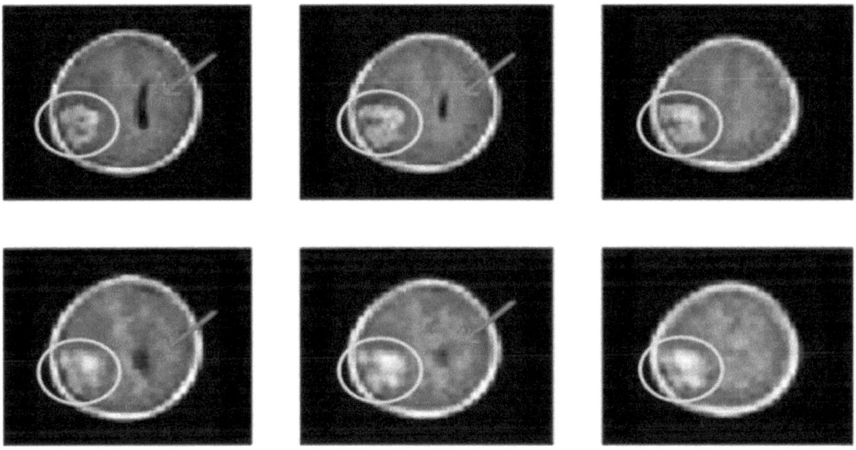

FIGURE 7.8 Brain tumor detection from MR image first row shows actual MR image and second row shows the processed images.

Source: Reprinted with permission from Ref. [16]. Copyright 2020. Elsevier.

7.14 APPLICATION OF COMPUTER VISION IN MEDICAL TRAINING

Computer vision is frequently utilized in medicine, not only for diagnosis but also for skill training. Currently, surgeons are not only reliant on the conventional method of learning skills through hands-on experience in the operating room. Instead, platforms for surgery based on simulation have shown to be effective teaching and evaluation tools. Before entering the operating room, trainees can practice their surgical skills using surgical simulation. Before performing an operation, it enables them to get accurate feedback and evaluations on how they performed, giving them a greater understanding of patient care and safety [19].

The development of Augmented and Virtual Reality (AR and VR) has a great potential in healthcare industry. AR/VR can be implemented to educate the medical students at early stage for training specific surgery. These technologies able to make the students learn interactively with the surrounding. Earlier, the medical students were difficult to understand various concepts of human body or some surgical procedure because of limited real-life experience by students. The AR and VR technology could create a virtual environment for medical procedure hence the learning experience of students has been enhanced. The medical students can learn about more complicated surgical procedure and anatomy of human without involving real patient for training [20].

7.15 APPLICATION OF COMPUTER VISION IN COMBATING COVID-19

The COVID-19 outbreak has posed a serious threat to the world's healthcare system. Computer vision can significantly aid in solving this challenge as nations all around the world work to combat the disease. Due to quick technological advancements, computer vision applications can assist in the diagnosis, control, treatment, and prevention of COVID-19. Digital chest X-ray radiography images combined with computer vision tools like COVID-Net make it simple to spot disease in patients. The prototype algorithm, created by Darwin AI in Canada, has produced COVID diagnostic results with an accuracy rate of 92.4%. Computer vision is used to do masked face identification, which is routinely employed to enforce and monitor pandemic disease prevention measures [21, 22].

7.16 APPLICATION OF COMPUTER VISION IN HOME-BASED PATIENT REHABILITATION AND MONITORING

Many patients prefer to recuperate at home instead of in a hospital after an illness. Medical practitioners can use computer vision programs to assist patients with appropriate physical therapy while also tracking their development electronically. This type of home instruction is not only more convenient, but also more cost-effective. Computer vision technologies can also help with non-intrusive remote monitoring of patients or the elderly. The goal of deep learning-based human fall detection systems, which are widely explored, is to lessen dependency and care expenses in the senior population. Another method of measuring the amount of time required to conduct a quick evaluation test to determine functional mobility is the video-assisted analysis of standardized medical exams like the TUG test. The TUG test can be used to gauge a person's likelihood of falling and their capacity for balance while walking [23, 24].

7.17 3D PRINTING

Invention of 3D printing technology has opened a new doorway of modeling in organs. With the availability of a CT or MRI scan image of any internal organ, with few hours, it can be printed into physical 3D object. The 3D printed model of internal organ can be useful for surgeons to understand the complex geometry of organ and effectively plan the surgery. The 3D bio-printing workflow of biomedical object illustrated in Figure 7.9 involves the following steps:

- Obtaining images of specific organ through CT or MRI scan;
- Segmentation and extracting the surface of organ;
- Converting the image data into 3D CAD based model;
- Converting the CAD model into STL file for 3D printing;
- Slicing and tool path generation for 3D printing;
- 3D printing and post-processing.

The most important and complex step in 3D bio-printing is conversion of CT or MRI image data into 3D CAD model. Various software packages were available for conversion of medical image to 3D model which process the images obtained from scan and construct the exact 3D model of that organ [25, 26].

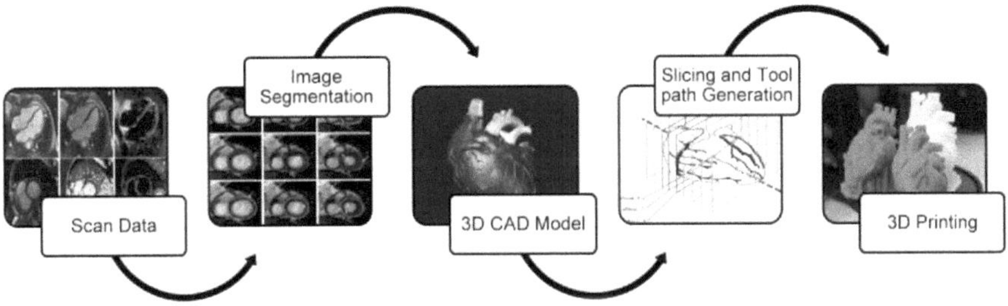

FIGURE 7.9 Workflow of 3D printing of heart.

7.18 CONCLUSION

Computer Vision holds a lot of promise for enhancing medical treatments all across the world. Computer vision applications significantly reduce the time and effort required to diagnose health issues or the effects of medication on a specific population. Many researchers are continually proposing new algorithms for processing and extracting features from medical image. The improvement in image processing techniques significantly increase its contribution in health-care field in assisting the doctors with more information for diagnosis. In near future, autonomous robot surgeon is possible which can lower medical error due to fatigue. The development of AR and VR technology is boon medical field, which is capable of educating the students with limited resources by simulating the human anatomical body. Still the AR and VR technology is at premature stage and more insights are needed to fully rely on this technology.

KEYWORDS

- **cesium iodide**
- **computed tomography**
- **flat panel detector**
- **positron emission tomography**
- **single-photon emission computed tomography**
- **thin-film transistor**

REFERENCES

1. Arena, P., Basile, A., Bucolo, M., & Fortuna, L., (2003). Image processing for medical diagnosis using CNN. *Nucl. Instruments Methods Phys. Res. Sect. A Accel. Spectrometers, Detect. Assoc. Equip.*, *497*, 174–178.
2. Scholl, I., Aach, T., Deserno, T. M., & Kuhlen, T., (2011). Challenges of medical image processing. *Comput. Sci. Dev.*, *26*, 5–13.
3. Nandi, D., Ashour, A. S., Samanta, S., Chakraborty, S., Salem, M. A. M., & Dey, N., (2015). Principal component analysis in medical image processing: A study. *Int. J. Image Min.*, *1*, 65–86.
4. Jia, G., Huang, X., Tao, S., Zhang, X., Zhao, Y., Wang, H., He, J., et al., (2022). Artificial intelligence-based medical image segmentation for 3D printing and naked eye 3D visualization. *Intell. Med.*, *2*, 48–53.

5. Sonka, M., & Fitzpatrick, J. M., (2000). *Handbook of Medical Imaging: Medical Image Processing and Analysis* (Vol. 2). University of Iowa. ISBN: 081948119X.
6. Chowdhary, C. L., & Acharjya, D. P., (2020). Segmentation and feature extraction in medical imaging: A systematic review. *Procedia Comput. Sci.*, *167*, 26–36.
7. Abdallah, Y. M. Y., & Alqahtani, T., (2019). Research in medical imaging using image processing techniques. *Med. Imaging-Principles Appl.*
8. Domingues, I., Pereira, G., Martins, P., Duarte, H., Santos, J., & Abreu, P. H., (2020). Using deep learning techniques in medical imaging: A systematic review of applications on CT and PET. *Artif. Intell. Rev.*, *53*, 4093–4160.
9. Von, H. F., Böttger, S., Wulff, D., Hagenah, J., García-Vázquez, V., & Ipsen, S., (2021). Medical robotics for ultrasound imaging: Current systems and future trends. *Curr. Robot. Reports*, *2*, 55–71.
10. Lundervold, A. S., & Lundervold, A., (2019). An overview of deep learning in medical imaging focusing on MRI. *Z. Med. Phys.*, *29*, 102–127.
11. Leitgeb, R. A., & Baumann, B., (2018). Multimodal optical medical imaging concepts based on optical coherence tomography. *Front. Phys.*, *6*, 114.
12. Gao, J., Yang, Y., Lin, P., & Park, D. S., (2018). Computer vision in healthcare applications. *J. Healthc. Eng.*
13. Olveres, J., González, G., Torres, F., Moreno-Tagle, J. C., Carbajal-Degante, E., Valencia-Rodríguez, A., et al., (2021). What is new in computer vision and artificial intelligence in medical image analysis applications? *Quant. Imaging Med. Surg.*, *11*, 3830.
14. Nam, K. W., Park, J., Kim, I. Y., & Kim, K. G., (2012). Application of stereo-imaging technology to medical field. *Healthc. Inform. Res.*, *18*, 158–163.
15. Kitaguchi, D., Takeshita, N., Hasegawa, H., & Ito, M., (2022). Artificial intelligence-based computer vision in surgery: Recent advances and future perspectives. *Ann. Gastroenterol. Surg.*, *6*, 29–36.
16. Hashemzehi, R., Mahdavi, S. J. S., Kheirabadi, M., & Kamel, S. R., (2020). Detection of brain tumors from MRI images base on deep learning using hybrid model CNN and NADE. *Biocybern. Biomed. Eng.*, *40*, 1225–1232.
17. Yunchao, G., & Jiayao, Y., (2019). Application of computer vision and deep learning in breast cancer assisted diagnosis. In: *Proceedings of the Proceedings of the 3rd International Conference on Machine Learning and Soft Computing* (pp. 186–191).
18. Bao, Y., Tang, Z., Li, H., & Zhang, Y., (2019). Computer vision and deep learning-based data anomaly detection method for structural health monitoring. *Struct. Heal. Monit.*, *18*, 401–421.
19. Seo, J., Han, S., Lee, S., & Kim, H., (2015). Computer vision techniques for construction safety and health monitoring. *Adv. Eng. Informatics*, *29*, 239–251.
20. Tang, Y. M., Ng, G. W. Y., Chia, N. H., So, E. H. K., Wu, C. H., & Ip, W. H., (2021). Application of virtual reality (VR) technology for medical practitioners in type and screen (T&S) training. *J. Comput. Assist. Learn.*, *37*, 359–369.
21. Ozturk, S., (2020). The COVID-19 from neurological overview/Norolojik Bakis Acisindan COVID-19. *Turkish J. Neurol.*, *26*, 56–58.
22. Ulhaq, A., Khan, A., Gomes, D., & Paul, M., (2020). *Computer Vision for COVID-19 Control: A Survey.* arXiv Prepr. arXiv2004.09420.
23. Gu, Y., Pandit, S., Saraee, E., Nordahl, T., Ellis, T., & Betke, M., (2019). Home-based physical therapy with an interactive computer vision system. In: *Proceedings of the Proceedings of the IEEE/CVF International Conference on Computer Vision Workshops.*

24. Ferreira, A. M., Salim, R., Fogagnolo, F., De Oliveira, L. F. L., Riberto, M., & Kfuri, M., (2021). The value of a standardized knee functional assessment in predicting the outcomes of total knee arthroplasty. *J. Knee Surg.*
25. Ozbolat, I., & Gudapati, H., (2016). A review on design for bioprinting. *Bioprinting*, *3*, 1–14. doi: 10.1016/j.bprint.2016.11.001.
26. Hill, E. R., Robertson, J. L., Maneas, E., & Bu, T. M., (2017). From medical imaging data to 3D-printed anatomical models. *PLoS One*, *12*, 1–10.

CHAPTER 8

ACTUATORS IN MEDICAL DEVICES

S. SASIKUMAR, S. SANTHAKUMAR, R. JAYAPAL, and R. THANIGAIVELAN

Department of Mechanical Engineering, Muthayammal Engineering College, Rasipuram, Namakkal, Tamil Nadu, India

ABSTRACT

This chapter provides a brief overview of the benefits and uses of actuators in MIS. Robot actuator design is determined mostly by the actuators' interconnections and degree of motion. Medical devices employ a range of actuators, including hybrid electromagnetic, tendon-driven, elastic fluidic, piston-cylinder, SMA, and DC micro-motors, to ensure optimal performance. Additionally, dental professionals employ piezoelectric actuators to grind down teeth and remove tartar. Actuators have a significant impact on patient room automation and rehabilitation devices, particularly when it comes to transferring immobile patients through the integration of the internet of things.

8.1 INTRODUCTION

Micro-actuators find application in various fields such as medical, HVAC, and aerospace industries. It is a micro-device able to generate mechanical motion to drive fluids and other masses. Micromotion is generated through converting electrical energy into kinetic energy. The size of the micro-actuators varies between few millimeters' "mm" to "μm" micrometers. The force developed by these actuators are in the range of nanoNewton to milliNewton range

Robotics and Automation in Healthcare: Advanced Applications. R. Thanigaivelan, Sanjay Singh, & Clement Christy Deepak C. (Eds.)

and flow rates in microfluidic actuators are generally indicated in pl/min to μl/min, respectively (pl = picoLiter and microLiter). Piezoelectric actuators play the prominent role in the construction of micro-actuators. This chapter discusses the various types of actuators and its applications.

8.2 ACTUATORS IN MINIMAL INVASIVE SURGERY (MIS)

The pneumatic and electrical rotary actuator plays a significant role in MIS. The outstanding benefits of MIS are reduction of postoperative pain, fewer complications during operation and post-operation, reduced hospital stay, faster recovery, less cut and scarring, less stress on the immune system, moreover, reduced operating time and costs. The commonly used devices such as graspers, forceps, electrode, baskets, etc., are considered in MIS. Katsuhiko Fukushima et al. [1] have proposed an actuator comprises a soft actuator and a cylindrical cam mechanism that exchanges linear into rotary movement. In Laparoscopic surgery, small rotary actuator was fitted on the edge of a forceps manipulator which produced 70° rotational angle and 3.0 N mm torque. Kim et al. developed a forceps manipulator with rigid link joints with multi-degrees of freedom (DOF). This manipulator inserts a four-DOF joint mechanism into patient body and experience a wide workspace.

8.2.1 FORCEPS MANIPULATOR

Forceps are miniature tool used to handle small tissues and dental parts. Suzuki et al. [3] have designed the compact forceps for assisting laparoscopic surgery. This type of forceps has rotational, translational, and pivoting mechanism. The work volume of this forceps is achieved with the help of friction wheel mechanism as 4,83,600 mm^3 with 0.6 kg weight and gimbals mechanism give pivoting motion around slited hole on the abdomen. Yamashita et al. [4] have fabricated a forceps manipulator for removing cells, holding, grasping, and stitching tissues. They developed 2-DOFs with +90° bending per DOF and 1-DOF forceps with less 1.0° bending and power generation of 0.4 kgf for lifting 0.5 kg of weight. Nobuhiko Miyata et al. [5] developed a micro-grasping forceps manipulator (Figure 8.1) with multi-functionality namely, 2° of bending freedom, precise positioning, and MR-compatibility.

Ishii & Kobayashi [6] have proposed a new robotic forceps manipulator with double-screw-drive (DSD) mechanism. This mechanism enables multi-directional bending by rotating two linkages comprises of right-handed and left-handed screw, universal joint, so that gripper can rotate. Moreover, the opening and closing motions of a gripper is achieved by wire-drive. Figure 8.2 shows the DSD manipulator with bending-rotation coordinates.

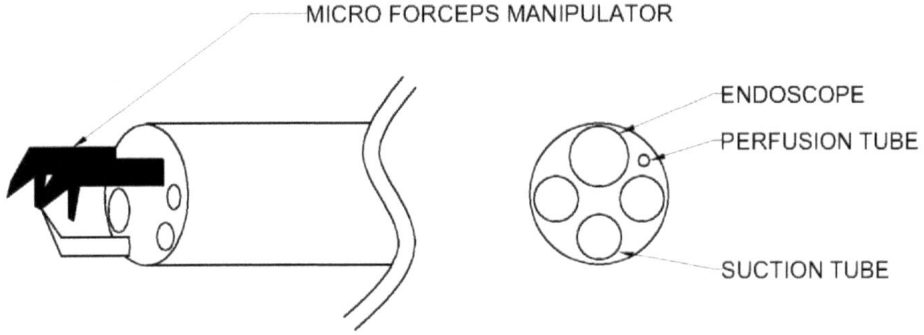

FIGURE 8.1 Multi-functionality forceps.

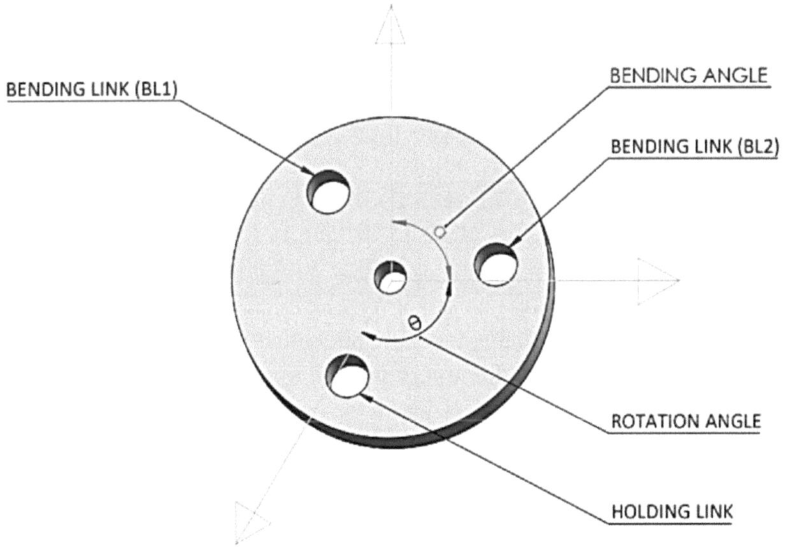

FIGURE 8.2 Motion coordinates [6].

8.3 TYPES OF ACTUATORS IN MIS

There are variety of actuators has been applied to MIS based on the need and operations. Hybrid electromagnetic type actuator was developed by Hao Chen Wang et al. [8] as tendon-driven type. It is a crucial device of MIS surgical tools to offer them with optimum angulation. Fischer et al. [7] have used shape memory alloys-spring to help the surgeon's fingers to feel the finger pressure distribution while using laparoscopic forceps. Other types of actuators namely elastic fluidic actuators and piston-cylinder actuators were also considered in MIS. Elastic fluidic actuators create lesser forces with minor displacements; nevertheless, piston-cylinder solutions need seals, which can create friction and involve maintenance costs for a good reliability [9]. Shape memory Alloys (SMA) plays the important role in design of actuators. Figure 8.3 show the SMA designed for MIS purpose. Kode & Cavusoglu [10] have developed an actuator by linking micro-motor (DC type) and a SMA actuator in series. This offers 1 mm stroke length with 8 mm jaws gap results 45° opening of driver jaw. A specific type of soft robotic actuators energized by a pressurized fluid (gas or liquid) specified as elastic inflatable actuators exhibit large strokes, very little friction and generates distributed force due to Pascal's principle [11]. Edinger et al. [12] have studied the use of piezoelectric actuator for MIS. This end-effectors and PZT inchworm actuator are part of a telerobotic system, which is planned to fit in both position control of the tool and force sensing. Tadano & Kawashima [13] have driven the slave manipulator with pneumatic actuators, which is having 7 DOF for use in MIS. Takemura et al. [14] have constructed single unit of multi-DOF ultrasonic actuator for creating multi-DOF rotation of a spherical rotor by using three natural vibration modes of a bar-shaped stator for MIS application. Ogura et al. [15] have used micro-rubber pneumatic actuators especially in MIS and handling biological cells because of high safety caused by low mechanical impedance. Sajid Nisar et al. [16] have developed the optimal configuration mechanism which satisfies the MIS workspace and achieves the maximum kinematic performance. Furthermore, this mechanism design is ideal for fully back drivable motion. This geometric design of the mechanism is simpler than the existing 2 DOFs and it uses fewer numbers of links and joints. Ottermo et al. [17] designed a laparoscopic grasper with of 32 micro-brushless motor. Banerjee et al. [18] have used hydrogel–polymer-based soft actuators in which the hydrogel material has the capacity to swell/shrink based on the quantity of water present, which grant us to create various movement and actuations. This water quantity is responsive for numerous external stimuli, namely magnetism, heat, electricity, ionic strength, etc., hydrogel-based actuators find huge applications in various fields even though it is having working limitations in aqueous media. The magnetic actuators are one type of actuator finds application for its advantages such as it can be rotary or linear, continuous or limited motion and multiple degrees-of-freedom of controlled motion.

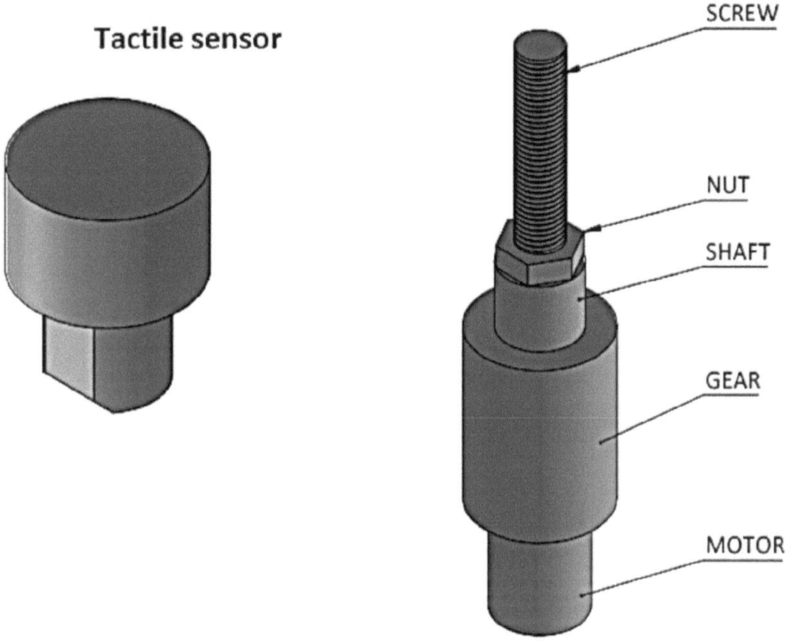

FIGURE 8.3 Motor in MIS [17].

Haptic technology helps the doctors to understand the amount of force they apply during surgery. Transparent actuator is used to create vibrations for human perception. An electroactive polymer is considered for developing flexible/deformable next-generation devices. Usually, haptic sensations are formed in the form of vibrations which will be varying based on the vibration amplitude and frequency. Two transparent conductive layers was included in haptic actuator (Layer with soft ionic). The dielectric layer is packed in between the two ionic conductive layers and is compacted in the thickness direction. In human perception, the application of electric field to the two ionic conductive layers it creates a vibration which is sufficiently strong. When the supply of electric field is made disconnected, the haptic actuators restored to their initial configuration. There are, three different dielectric elastomers namely, silicone rubber, polyurethane, and acrylic polymer are considered for the dielectric layer. The haptic performance of these vibrotactile actuators were studied using an accelerometer. Among the different elastomers silicone rubber-based actuator, at the resonant frequency (80 Hz) the measured acceleration is 1.408 g (g = 9.8 m/s^2) and the response time is about 2.8 ms [22]. Kim et al. [23] have developed transparent and stretchable graphene-based actuator for superior tactile displays. Here the actuator consists of transparent graphene electrodes and also dielectric elastomer substrate (DES). In this the electrode the substrate is coated in layer by layer on the selected region, which will bump up the response to the input voltage, which accordingly brings actuation.

Dielectric elastomer actuators have been applied to many areas such as soft robots, artificial muscles, and compliant devices. Wang et al. [25] have developed transparent artificial muscles by using blends of poly(3,4-ethylene dioxythiophene), poly(styrene sulfonate) and waterborne polyurethane (WPU) as compliant electrodes. The poly(styrene sulfonate) blends are conductive and translucent solids, and the dielectric elastomer actuator artificial muscles can attain a voltage-induced area strain of 200% with a transmittance of higher than 88% over the whole observable light spectrum. Biomimetic applications demand large deforming soft actuators with high stability in response to multi-stimuli. The limitations in current actuators like low performance and high energy consumption have forced the researchers to work on sandwich structural soft actuator (MPDMS/MXene/PTFE) which possesses both electrothermal/magnetic actuations. The soft actuator shows a good electrothermal actuation and the bending deformation (353°C) due to its large difference of thermal expansion coefficient among MPDMS film and PTFE tape at 3 V power supply. Sooner or later, the electrothermal/magnetic coupling actuation can be applied in an intelligent crawling robot to overcome the conventional inchworm movement simulation [26]. In use of soft actuators, the position sensing is major limitation in the existing system which can be reduced with facilitation of closed loop control. A pneumatic actuator (bending type) with an incorporated carbon nanotube position (ICNP) sensor was assembled. This ICNP sensor measure the large displacements regularly come across in soft robotics. It is formed by a simple soft lithography process. Calderón et al. [27] have designed, fabricated and feedback control for earthworm-inspired multi-material multi-actuator soft robot competent of locomoting inside pipes. The body of natural earthworms comprises the repeated deformable structural units, which generate the peristaltic body motions vital for limbless underground burrowing and ground crawling. As in earthworm, a new pneumatically driven soft robotic system replicates the action of a single burrowing earthworm's segment. The horizontal, vertical, and oblique movement test was performed in transparent pipe to ensure real time usage and demonstration.

8.4 ACTUATORS IN DENTISTRY

In dentistry, actuators have indispensable space and Yokoyama-Sato et al. [28] have developed the actuator-driven pulsed water jet which have ability to reduce water consumption and selectively remove tissues. The setup enables soft tumor removal with protection of the nerves and blood vessels. The actuators also find application in 3D intraoral scanning apparatus. Piezo stage and a liquid lens were used to reduce the mechanical motion of optics inside the 3D dental scanning apparatus by replacing the motor-driven actuators. In addition, the use of piezo stage in scanning apparatus makes the apparatus more compact [29]. The haptic system is adapted in dentistry which comprises of input device, a processor, and an actuator. Haptics is created by a surgeon which triggers the processor to turn on an actuator that make a feedback event such as vibration. In haptics aided operative procedure a blend of digital information and computer assisted tactile feedback could lessen the incidence of unavoidable mistakes by guiding the Dentist while performing cavity preps [30]. A piezoelectric ultrasonic bone surgical instrument generally used to take out the tartar out of teeth. It is an accepted instrument for dental treatment for its advantages such as small size, low-electric power, and precision control of surgical operation. This instrument consists of two parts of a tip and vibration system which is also includes head, piezoelectric elements, and tail-mass [31]. Pecheva et al. [32] have assessed the vibration motion of a piezoelectric scaler for better entry into periodontal pockets.

8.5 CUSTOM REHABILITATION DEVICES

The majority of the robotic rehabilitation devices are actuated by electric motors, which provide enough power and perfect motion. At the same time, it has drawbacks namely weight and restricted direction of motion which reduce its importance in rehabilitation devices. On comparing with heavy actuators, soft actuators, have better qualities due to their compliance and flexibility. Soft wearable devices manufacturing mainly focuses on actuators. The Soft wearable devices actuation is done by fluid pressurization or cables. Traditionally for fluid pressurization, pneumatic actuation is used for inflatable structures, pneumatic artificial muscles and fluid elastomer actuators which also use hydraulic energy. With respect to cable actuation, the diversified design like dielectric elastomer actuators, electro active polymers,

SMA or tendon and muscle wires are considered. Wang et al. [33] have developed finger rehabilitation robot which comprises mechanical structure, driver, transmission, and actuating mechanisms, and the magnetorheological damping device. This mechanism accurately transfers the driving force to the joints of the finger for the autonomous training of every joint while minimizing the device weight. Pylatiuk et al. [34] have developed elbow orthosis by integrating electromyographic, functional electrical stimulation, and micro-flexible fluidic actuators. This type of elbow orthosis generates 6.5 Nm flexion moments against gravity, while 3.1 Nm are necessary to hold an average forearm against gravity and a maximum of 5.8 Nm are exerted during daily activities. In powered orthotics most important technical challenges are energy storage capacity, compact rotary or linear actuator and packaging of components [35]. Human motion is associated with relatively high torque and low velocity actuation. But, electric motors (DC) generate low torque, high velocity, moreover electric motors need to transmit rotary to linear motion through ball screw, to match the actuator to the load requirements. Fluid power system of actuation has uniqueness that high force-to weight and force-to-volume ratios. The pressurized fluid is sent to the actuator through flexible hoses in which space requirements are reduced. While considering tiny, low power hydraulic systems, the friction losses and drag losses in cylinders, pumps, and in conduits go up due to surface area to volume scaling. William Durfee et al. have focused on development of untethered orthotics with small-scale hydraulics and analysis indicates a clear opportunity for applying tiny hydraulics to untethered powered orthotics.

8.6 OTHER APPLICATIONS OF ACTUATORS

Actuators also find application in the patient room for handling the passive and partially disable patient. Endo et al. [36] have developed the multitasking robot for patient assistance. The main components are a multimodal man-machine interface, a server, and a multi-directional mobile robot. The system monitors the movement of the patient needs care support and decides the essential help. The actuator in the robot includes a motor-controlled mechanism which transfers rotational motion between the gears. Lui et al. [37] have developed the hospital bed for bed ridden patient. This bed consists of carbon fiber frames, eye movement tracker and actuators with positioning control. The eye gaze position is sensed to understand the intention of user and provide a special service that the patient expects. With the help of automation and intelligence via the technology from Internet of Things (IoTs) makes paradigm shift in healthcare industry by ensuring utmost patient care. Tech Lift of Spring City, Pa., have addressed the problem of back injury that results due to the poor handling of patient while lifting the patient on a stretcher. They developed a compressor powered self-actuating stretcher for patient transport which comprises the air spring. Compressed air supplied will expand the air spring to lift the stretcher, and by leasing the simple lever, the patient can be lifted. This type of arrangement offers a softened ride for patients, helps to reduce the chance of injury. Ireneusz Malujda et al. [39] have developed bed with kinematic pair turning around the z axis, scissors lifting system, guide sleeves and guide rails. The hydraulic actuator is the part of the bed with a foot powered pump enclosed in a single body. The maximum axial force exerted by this actuator is F = 10 kN, and the maximum stroke length of the rod is 300 mm. This type of mechanism is found suitable for bed deck height adjustment within the required range after analysis. The long bedridden patient is prone to bedsores due to extended pressure on the skin and core tissues. The bedsores are caused by pressure, shear, and friction when a person lies on a bed for a prolonged period of time. The pressure prone areas in the body are tailbone, blades in shoulders and ankles where the blood flow are restricted. To alleviate this problem the pressure is distributed throughout the body by varying the pressure points frequently. Premkumar et al. [40] have designed an automated bed with 4 linear actuators and a specifically designed three plate platform system. This bed is able to produce five different combinations of actuation namely sitting, left tilt, right tilt, Trendelenburg and reverse-Trendelenburg. For lumbar treatment new type of multi-dimensional bed were developed with two degrees of freedom to realize a traction treatment of lumbar through different motion such as flexion, extension, and rotation. Two linear actuators have been used to provide motion with high control accuracy. Spano and Asada from MIT [7] inspired by natural waves namely water waves that can take away objects, is used for moving a human body on the surface of a bed. The developed partial bed surface with series of slider-crank mechanisms creates a wave to move a body to the edge of the bed.

Seon et al. [43] have developed a robotic bed with multiple segments which are driven separately by brushless DC motors. The body pressure information was collected with the help of a belt-type body pressure sensor for feedback control. The movements of the segments on the top of the bed are controlled with a fuzzy controller. This bed integrates the main controller, a teach pendant, controllers, and sensors.

8.7 CONCLUSION

This chapter gives the brief description about the application of actuators in MIS and its advantages. The degree of motion and linkages of the actuators make the decisive for designing the robot actuators. Variety of actuators namely hybrid electromagnetic, tendon-driven, elastic fluidic actuators, piston-cylinder actuators, SMA and DC micro-motor are used in medical devices for efficient operation. Moreover, in Densitry piezoelectric actuators are used for removing tartar out of teeth and tooth grinding. Actuators plays impressive role in rehabilitation devices and patient room automation, especially for moving the bedridden patient's integration of internet of things plays a dominant role.

KEYWORDS

- **degrees of freedom**
- **dielectric elastomer substrate**
- **double-screw-drive**
- **minimal invasive surgery**
- **shape memory alloys**
- **waterborne polyurethane**

REFERENCES

1. Katsuhiko, F., Tetsuro, M., Toshihiro, K., Takahiro, K., Maina, S., & Yoshikazu, N., (2022). A pneumatic rotary actuator for forceps tip rotation. *Sensors, and Actuators A: Physical, 333*, 113222. doi.org/10.1016/j.sna.2021.113222.
2. Kim, K., Song, H., Suh, J., & Lee, J., (2010). A teleoperated minimally invasive surgical system with an additional degree of freedom manipulator. In: *Proceedings of the Fourth International Conference on Sensor Technologies and Applications* (pp. 90–94). doi: 10.1109/SENSORCOMM.2010.21.
3. Suzuki, T., Katayama, Y., Kobayashi, E., & Sakuma, I., (2005). Compact forceps manipulator using friction wheel mechanism and gimbals mechanism for laparoscopic surgery. In: *International Conference on Medical Image Computing and Computer-Assisted Intervention* (pp. 81–88). Springer, Berlin, Heidelberg.
4. Yamashita, H., Kim, D., Hata, N., & Dohi, T., (2003). Multi-slider linkage mechanism for endoscopic forceps manipulator. In: *Proceedings 2003 IEEE/RSJ International*

Conference on Intelligent Robots and Systems (IROS 2003) (Cat. No. 03CH37453) (Vol. 3, pp. 2577–2582). IEEE.

5. Miyata, N., Kobayashi, E., Kim, D., Masamune, K., Sakuma, I., Yahagi, N., & Takakura, K., (2002). Micro-grasping forceps manipulator for MR-guided neurosurgery. In: *International Conference on Medical Image Computing and Computer-Assisted Intervention* (pp. 107–113). Springer, Berlin, Heidelberg.
6. Ishii, C., & Kobayashi, K., (2006). Development of a new robotic forceps manipulator for minimally invasive surgery and its control. In: *2006 SICE-ICASE International Joint Conference* (pp. 250–253). IEEE.
7. Fischer, H., Trapp, R., Schüle, L., & Hoffmann, B., (1997). Actuator array for use in minimally invasive surgery. *Journal de Physique IV Proceedings, EDP Sciences, 07*(C5), C5-609-C5-614. ff10.1051/jp4:1997596ff.ffjpa-00255696f.
8. HaoChen, W., SaiHui, C., Yao, W., & ChengLi, S., (2020). A hybrid electromagnetic and tendon-driven actuator for minimally invasive surgery. *Actuators*, *9*(3), 92. https://doi.org/10.3390/act9030092.
9. Paterno, L., Tortora, G., & Menciassi, A., (2018). Hybrid soft–rigid actuators for minimally invasive surgery. *Soft Robotics*, *5*(6), 783–799.
10. Kode, V. R. C., & Cavusoglu, M. C., (2007). Design and characterization of a novel hybrid actuator using shape memory alloy and dc micromotor for minimally invasive surgery applications. *IEEE/ASME Transactions on Mechatronics, 12*(4), 455–464.
11. Gorissen, B., Reynaerts, D., Konishi, S., Yoshida, K., Kim, J. W., & De Volder, M., (2017). Elastic inflatable actuators for soft robotic applications. *Advanced Materials*, *29*(43), 1604977.
12. Edinger, B., Frecker, M., & Gardner, J., (2000). Dynamic modeling of an innovative piezoelectric actuator for minimally invasive surgery. *Journal of Intelligent Material Systems and Structures*, *11*(10), 765–770.
13. Tadano, K., & Kawashima, K., (2010). Development of a master–slave system with force-sensing abilities using pneumatic actuators for laparoscopic surgery. *Advanced Robotics*, *24*(12), 1763–1783.
14. Takemura, K., Park, S., & Maeno, T., (2008). Control of multi-dof ultrasonic actuator for dexterous surgical instrument. *Journal of Sound and Vibration*, *311*(3–5), 652–666.
15. Ogura, K., Wakimoto, S., Suzumori, K., & Nishioka, Y., (2009). Micro-pneumatic curling actuator-nematode actuator. In: *2008 IEEE International Conference on Robotics and Biomimetics* (pp. 462–467). IEEE.
16. Nisar, S., Endo, T., & Matsuno, F., (2017). Design and kinematic optimization of a two degrees-of-freedom planar remote center of motion mechanism for minimally invasive surgery manipulators. *Journal of Mechanisms and Robotics*, *9*(3), 031013.
17. Ottermo, M. V., Stavdahl, O., & Johansen, T. A., (2005). Electromechanical design of a miniature tactile shape display for minimally invasive surgery. In: *First Joint Eurohaptics Conference and Symposium on Haptic Interfaces for Virtual Environment and Teleoperator Systems – World Haptics Conference* (pp. 561, 562). IEEE.
18. Banerjee, H., Suhail, M., & Ren, H., (2018). Hydrogel actuators and sensors for biomedical soft robots: Brief overview with impending challenges. *Biomimetics*, *3*(3), 15.
19. Sallé, D., Cepolina, F., & Bidaud, P., (2004). Surgery grippers for minimally invasive heart surgery. In: *IEEE International Conference on Intelligent Manipulation and Grasping (IMG)* (pp. 1, 2).

20. Howe, D., (2000). Magnetic actuators. *Sensors and Actuators A: Physical, 81*(1–3), 268–274.
21. De Lorenzo, D., De Momi, E., Dyagilev, I., Manganelli, R., Formaglio, A., Prattichizzo, D., & Ferrigno, G., (2011). Force feedback in a piezoelectric linear actuator for neurosurgery. *The International Journal of Medical Robotics and Computer Assisted Surgery*, *7*(3), 268–275.
22. Choi, D. S., Lee, S. H., & Kim, S. Y., (2020). Transparent and soft haptic actuator for interaction with flexible/deformable devices. *IEEE Access*, *8*, 170853–170861.
23. Kim, U., Kang, J., Lee, C., Kwon, H. Y., Hwang, S., Moon, H., & Choi, H. R., (2013). A transparent and stretchable graphene-based actuator for tactile display. *Nanotechnology*, *24*(14), 145501.
24. Zhao, H., Hussain, A. M., Israr, A., Vogt, D. M., Duduta, M., Clarke, D. R., & Wood, R. J., (2020). A wearable soft haptic communicator based on dielectric elastomer actuators. *Soft Robotics*, *7*(4), 451–461.
25. Wang, Y., Gupta, U., Zhu, J., Li, P., Du, D., Zhang, L., & Foo, C. C., (2019). Bio-inspired soft robot driven by transparent artificial muscle. In: *2019 IEEE International Conference on Robotics and Biomimetics (ROBIO)* (pp. 1959–1964). IEEE.
26. Li, W., Sang, M., Liu, S., Wang, B., Cao, X., Liu, G., & Xuan, S., (2022). Dual-mode biomimetic soft actuator with electrothermal and magneto-responsive performance. *Composites Part B: Engineering*, 109880.
27. Calderón, A. A., Ugalde, J. C., Zagal, J. C., & Pérez-Arancibia, N. O., (2016). Design, fabrication, and control of a multi-material-multi-actuator soft robot inspired by burrowing worms. In: *2016 IEEE International Conference on Robotics and Biomimetics (ROBIO)* (pp. 31–38). IEEE.
28. Yokoyama-Sato, Y., Nishioka, T., Naganuma, Y., Takahashi, M., Nakagawa, A., Yoda, N., & Iikubo, M., (2022). Effectiveness of an actuator-driven pulsed water jet for removal of softened carious dentin. *Dental Materials Journal*, 2021–2234.
29. Ahn, J. S., Park, A., Kim, J. W., Lee, B. H., & Eom, J. B., (2017). Development of three-dimensional dental scanning apparatus using structured illumination. *Sensors*, *17*(7), 1634.
30. Natarajan, M., (2018). An innovative application of haptics technology to reduce iatrogenic errors in operative dentistry–a novel idea. *Oral Health Dental Sci.*, *2*(3), 1–4.
31. Lee, J. H., Oh, J. M., Hong, Y., Kim, S., Paik, J., Lee, Y. J., & Lee, S. D., (2012). Design and evaluation of ultrasonic bone surgical instruments for dental application. *Journal of the Korean Institute of Electrical and Electronic Material Engineers*, *25*(12), 990–995.
32. Pecheva, E., Sammons, R. L., & Walmsley, A. D., (2016). The performance characteristics of a piezoelectric ultrasonic dental scaler. *Medical Engineering and Physics*, *38*(2), 199–203.
33. Wang, D., Wang, Y., Zi, B., Cao, Z., & Ding, H., (2020). Development of an active and passive finger rehabilitation robot using pneumatic muscle and magnetorheological damper. *Mechanism and Machine Theory*, *147*, 103762.
34. Pylatiuk, C., Kargov, A., Gaiser, I., Werner, T., Schulz, S., & Bretthauer, G., (2009). Design of a flexible fluidic actuation system for a hybrid elbow orthosis. In: *2009 IEEE International Conference on Rehabilitation Robotics* (pp. 167–171). IEEE.
35. Durfee, W., Xia, J., & Hsiao-Wecksler, E., (2011). Tiny hydraulics for powered orthotics. In: *2011 IEEE International Conference on Rehabilitation Robotics* (pp. 1–6). IEEE.

36. Endo, M., Endo, M., & Kakizaki, T., (2017). Development of a rotational meshing actuator for in-home life support systems. *ROBOMECH Journal*, *4*(1), 1–12.
37. Lui, P. W., Lai, F. M., Su, K. C., Lin, J. Y., Chi, H. W., Wang, J. S., & Chen, Y. W., (2017). Use eye tracker to design an intelligent patient bed. *Energy Procedia*, *143*, 553–558.
38. https://www.machinedesign.com/archive/article/21813465/airspring-actuator-lifts-medical-stretcher (accessed on 20 November 2023).
39. Malujda, I., Wilczyński, D., Talaśka, K., Wojtkowiak, D., Szulc, M., Vasko, M., Handrik, M., et al., (2018). Study of the prototype mechanism of height adjustment of the bed in hospital bed. *MATEC Web of Conferences*, *157*, 02028. doi: 10.1051/matecconf/201815702028.
40. Premkumar, S., Girithar, T. A., Kumar, P. N., Saimothish, R., Malani, V. A., & Jacob, R., (2021). 5-position 4-actuator automated bed to prevent bed sores. In: *Journal of Physics: Conference Series* (Vol. 1831, No. 1, p. 012024). IOP Publishing.
41. Luo, Y., Liu, L., Liu, Y., & Ma, J., (2019). Control system design of multi-dimensional lumbar traction treatment bed. *Vibroengineering Procedia*, *26*, 18–23.
42. Spano, J. S., & Asada, H. H., (2000). Kinematic analysis and design of surface wave distributed actuators with application to a powered bed for bedridden patients. *IEEE Trans. Robot. Autom.*, *16*(1), 1–11.
43. Seon, M., Lee, Y., & Moon, C., (2021). Medical robotic bed to prevent pressure sores. *Applied Sciences*, *11*(18), 8459.

CHAPTER 9

SMART SPECTACLES WITH EMBEDDED SEIZURE DETECTOR

M. SOUNDARRAJAN, J. PREETHA, M. ARAVINDH, and M. VISHNU

Muthayammal Engineering College, Rasipuram, Namakkal, Tamil Nadu, India

ABSTRACT

IoT has become more popular in recent times; it can convert any normal device into a smart device. This concept is about using IoT enabled glasses for detecting and alerting at the time of seizure also known as fits or convulsion. The detection is based on two main parameters of the human body, which are body temperature and blood pressure. The blood pressure sensors and temperature sensors are placed in the temples of the glasses. There are minimum and maximum threshold limits for both temperature and blood pressure, if the value obtained from the sensors exceeds the limit or goes below the limit then the patient has higher changes of getting convulsion sooner. The sensors are connected to an IoT enabled device which constantly monitors the values for any abnormalities in the blood pressure or body temperature. If the device detects any abnormalities, then it immediately alerts the patient through the buzzer which indicates that they have high chances of getting convulsion sooner and also send an SMS to the registered mobile numbers about the condition of the patient with their location.

9.1 INTRODUCTION

The info-graph shown in Figure 9.1 depicts that 80% of the people with seizure are living in the developing countries. Currently, the ratio of people

Robotics and Automation in Healthcare: Advanced Applications. R. Thanigaivelan, Sanjay Singh, & Clement Christy Deepak C. (Eds.)

with active epilepsy is from 4 to 10 per 100 people. Each year, approximately 5 million people are diagnosed with epilepsy worldwide [18].

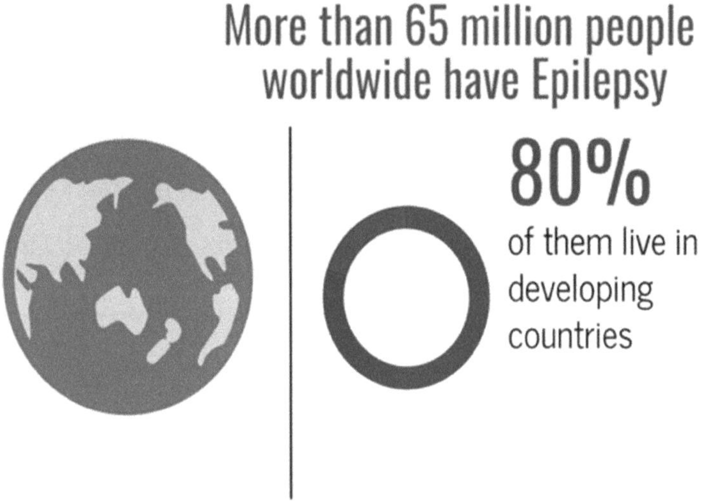

FIGURE 9.1 Percentage of people with seizures living in developing countries [18].

Gregg & Crepeau [1] reviewed about the usage of medical device for monitoring and identifying seizures and transmit the information to the caregiver alerts. They reviewed the generators and biosignals considered for seizure detection. Multimodal seizure recognition systems merge the multiple biosignals into a single system in the view of improving sensitivity of seizure detection. Kramer et al. [2] have developed an epilepsy detection system with motion sensor mounted on a wrist bracelet. The accelerometer has capacity to detect three-axis acceleration and transmit the information through the transmitter to the computer having algorithm to examine the recorded data. Seizures' alerts were matched up to with the video-EEG data [2]. The algorithm is able to identify 91% seizures and generate an alarm within a time of 17 seconds and system failed to identify 2 of 22 seizures (9%). Krishnu et al. [3] have developed an alert system that can facilitate caretakers to reach the patients fast. It consists of cloud services that receive signal from the user and record it. Urgent situation call is activated through a raspberry pi–GSM module interface [4].

A mobile application shares the location details to caretakers as google map link and address through call and message The average delays for calls and message received at caretaker's end are recorded as 13.6 and 8 seconds, respectively [3].

9.1.1 SEIZURE AND SYMPTOMS

Fits are unexpected, unrestrained electrical disorder in the brain. It can result in unusual change in patient behavior, movements, and degree of consciousness. Fits less than 2 times is normal but if it exceeds more than 2 times or the patient having recurrent fits then the condition is known as epilepsy. There are many classifications of fits based on their severity. The part of the brain from where the seizure originates decides the nature of the seizure [4]. The average time duration of fits ranges from 30 seconds to 2 minutes but if the fits last for a longer time like 5 minutes, then this condition should be considered critical and immediate treatment is required. Fits are not uncommon as you might think many people encounter fits in their life. Fits can occur after a stroke, a severe head injury, due to some infections like meningitis, etc. (Figure 9.2).

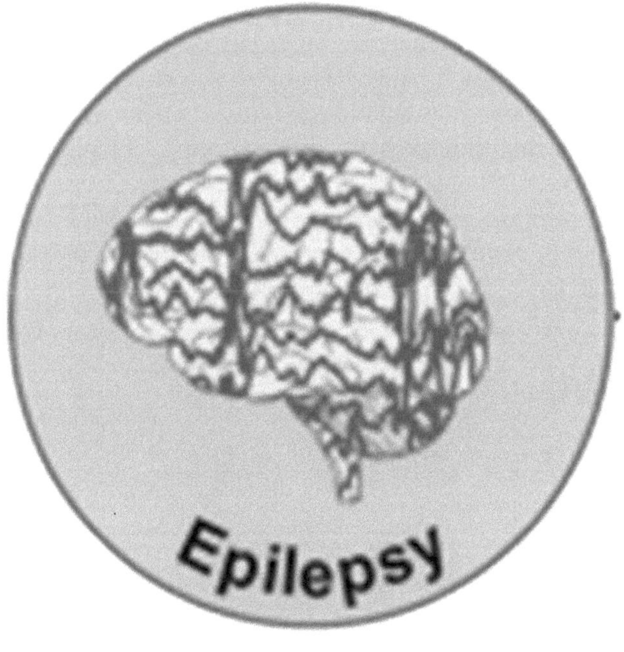

FIGURE 9.2 Effect of seizure on brain [4].

The symptom of seizure generally varies from the type and nature of the seizure the person is affected with. Some common symptoms of seizure are:

- Feeling sudden confusion for no reason;
- A staring spell;
- Legs and arms move unwillingly without control;
- Loss of consciousness or awareness;
- Feeling frightened or some form of anxiety.

9.2 TYPES OF SEIZURES

A type of seizure which is triggered due to unusual electrical activity in any area of brain is known as focal seizure. Generally, these seizures make changes to their consciousness and awareness which usually give them the feeling of being in dream. Even though the patients may seem awake, they will not respond to us and make some uncontrollable jerking movements. These may include rubbing their hands, abnormal mouth postures, talking repeatedly or walking in a circular pattern, etc. Sometimes, the patients do even remember that the seizure had occurred to them. Other type of seizures may trick their emotions or tricks other sense organs and give false information, but the patient doesn't lose consciousness. It may give them random feelings like happiness, trauma, angry or joyfulness. Some people reported that they've got a different kind of feeling which cannot even be described. People with these seizures also have conditions like difficulty in speaking, sudden jerking of body parts, such as arms or legs, and other kind of disturbing symptoms like tingling, dizziness, and seeing flashlights [19].

9.2.1 GENERALIZED SEIZURES

There are many types of seizures, which include:

- Absence seizures;
- Tonic seizures;
- Atonic seizures;
- Clonic seizures;
- Myoclonic seizures;
- Tonic-clonic seizures.

9.2.2 CHANGES IN BODY BEFORE SEIZURE

There are certain changes in the human body right before the seizure shows up. Two of the main changes are in blood pressure and body temperature. The sudden increase or decrease in blood pressure and temperature indicates that the person has higher chances of getting convulsion sooner. The normal blood pressure ranges from 90/60 mmHg to 120/80 mmHg and the average temperature ranges from 97°F (36.1°C) to 99°F (37.2°C) [3]. A person likely to have seizure sooner may have a blood pressure below 90/60 mmHg or above 120/80 mmHg and temperature below 95°F or above 100.4°F.

9.2.3 INTERNET OF THINGS (IOT)

The internet of things system is an interconnected computing devices with each assigned a unique identifier and has the ability to send and receive data through internet automatically without manual assistance.

In simple words it is the conversion of normal devices into smart devices with the help of internet and network enabled peripherals.

In past few years organizations of different categories use IoT to operate more effectively, monitor their workflow efficiently, improved statistical analysis and much more.

9.2.4 WORKING OF IOT

An IoT system consists of a network enabled smart peripherals that use embedded devices like processors, network interfacing hardware and various type of sensors to fetch and process the data they acquire from their connected devices. Then the device shares the data they've processed to an IoT gateway or any proprietary server devices where it is further processed locally or sent to cloud for processing. These devices are also capable of communicating with each other related devices and process on the combined data of all devices to yield the final result data. Most of the IoT devices are intended to be automatic however the initial setup needs some manual implementation [5].

9.2.5 IOT IN MONITORING PEOPLE WITH SEIZURE

Seizure is an uncontrollable as well as unpredictable disease, which asserts that the patients need to be monitored constantly. Our objective is to monitor the seizure patient with the help of special spectacles which consists of a temperature sensor with blood pressure sensor connected to an IoT enabled device which constantly monitors the patient for any abnormalities in temperature and blood pressure. If the device detects any abnormalities, then it immediately notifies the patient that they have higher chances of getting seizure sooner and also sends SMS to the registered mobile numbers about the condition of the patient with their current location [2].

9.2.6 *FOCUSING ON HEALTHCARE*

As the technology is growing more and more it keeps focusing in making the day-to-day tasks easier, automation, and entertainment but it doesn't care about its side effects on humans' health. A healthy living must be the highest priority of a human being. Efficient use of technology will be achieved only if it is useful for maintaining a good healthcare.

9.3 DEVELOPMENT OF THE PROTOTYPE

Prototype Specifications and Cost:	
1x Spectacles	$16.41
1x Optical heart-rate sensor	$6.31
1x Thermostat temperature controller	$2.52
1x Arduino board	$12.62
1x SIM8001 GSM GPRS module	$3.38
3x 9.0 V battery	$2.52
Others	$2.00
Total	**$45.76 (INR: 3,600)**

The temperature sensor and optical heart-rate sensors are placed in adjacent sides of the temples of the spectacles which constantly send the values of the temperature and blood pressure to the Arduino board where the data will be further processed. The sensors are placed on the temples and connected to the board in this prototype, but the final concept is to embed the sensors into the temples, i.e., the temples will act as sensors and the data derived from the sensors will be transferred wirelessly to the device (Figure 9.3).

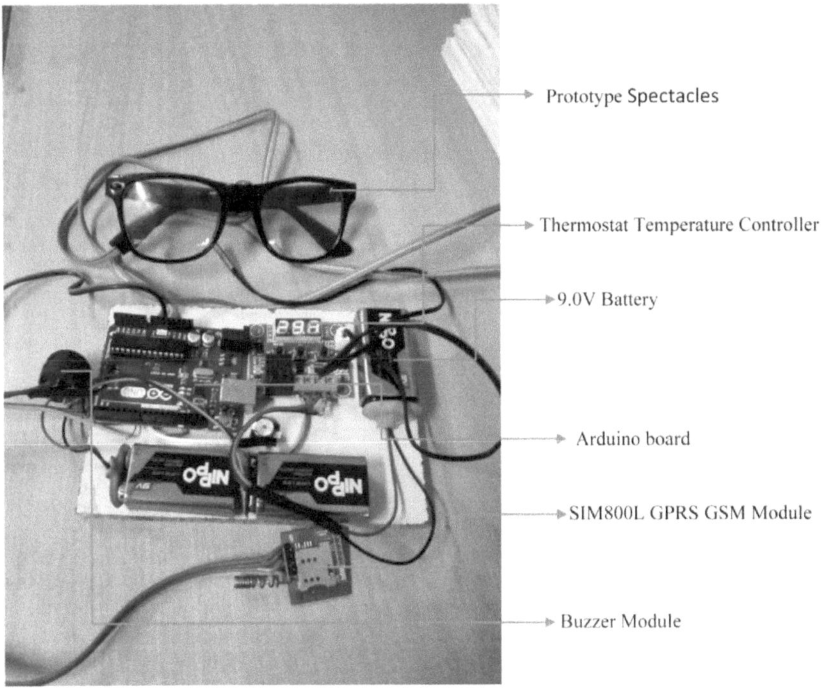

FIGURE 9.3 Components used in the project.

9.3.1 PROTOTYPE SPECTACLES (Figure 9.4)

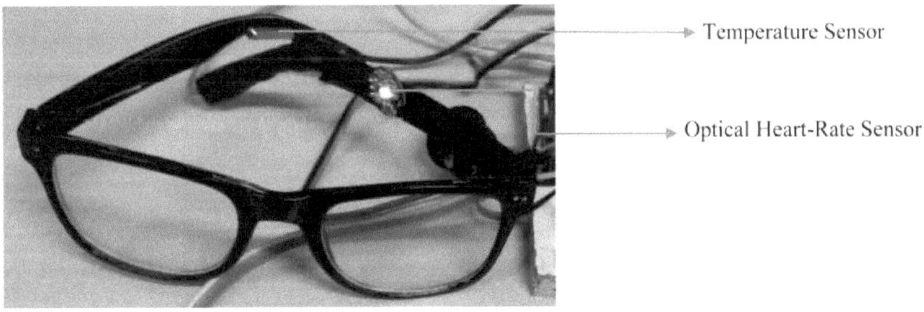

FIGURE 9.4 The spectacles with sensors on both sides.

9.3.2 THERMOSTAT TEMPERATURE CONTROLLER MODULE

Temperature sensors use electrical signals to measure the temperature readings. It has two metal structure that produces electrical voltage or resistance whenever a temperature change occurs. The digital temperature controller module is used to display the temperature readings obtained from the sensor. For the purpose of proof of concept, the display module is used in this project (Figure 9.5). This module requires a 12 V power supply to function.

FIGURE 9.5 W1209 thermostat module.

A W1209 thermostat module consists of the following components:

- Temperature sensor;
- Keys;
- LED display;
- Relays.

Thermostat devices is used to maintain temperature of a system at a particular point and automatically adjusts by sensing the temperature change. NTC temperature sensor uses intelligent control system to control various electrical devices of different temperatures [20].

Whenever the temperature increases, the resistance decreases in the NTC module. Since W1209 module has its own built-in micro-controller chipset manual programming is not necessary in most cases. The module has 3 switches for control, including the power on/off switches. The relay can operate at 240 V AC at 5 A or 14 V DC at 10 A. The Unit of temperature measured is in Degree Centigrade that is displayed using the 7-segment display array [20].

➢ **Specifications:**
- Temperature range supported – 50°C to 110C;

- The resolution ranges from – 9.9°C to 99.9C;
- Accuracy of measurement is at 0.1°C;
- The accuracy of control is at 0.1°C;
- The temperature reading is periodically refreshed for every 0.5 seconds.

The DC input range is from 9 V to 12 V and exceeding the limit of 12 V can cause the device to malfunction or even failure. While using it in places such as automobiles the battery voltage will be far higher than 12 V so a regulator or series of resistors must be used to limit the power supply to 12 V.

- Measuring input: NTC (10 K 0.5%);
- This sensor is waterproof and can work inside water up to a depth of 0.5 M;
- The output of relay 1 is 20 A (max).

9.3.3 ARDUINO BOARD

Arduino is an extraordinary platform for the construction of various kind of electronic projects. Arduino is a combination of physical hardware and set of software's for programming the board that runs on the user's pc to upload and write computer code to the Arduino board. The Arduino is a beginner friendly board, so it had gained more popularity among the beginners.

In most of the other predecessor hardware's a separate hardware is required for loading new code into the board but that's not the case in Arduino. The users can user their USB cable for writing the code in the board. The programming used in the Arduino board is kind of a more sophisticated and simplified version of C++ so that users who are experienced in C language can easily adapt to Arduino Programming. Arduino provides package-based modules for simplifying lot of tasks [21] (Figure 9.6).

The Arduino board is the main controller of the device, which is responsible for handling the input data from the sensors placed in the device. The code checks if the temperature and blood pressure value exceed the threshold limit or go below the limit. If so, then the Arduino sends an alert message to the user itself that they have higher chance or getting a seizure sooner and also sends an SMS to all the registered mobile numbers with the help of SIM module attached with this device [5].

FIGURE 9.6 Arduino board.

9.3.4 SIM800L GPRS GSM MODULE

Quad-Band GSM is supported in this module, we can use GPRS, SMS, call, etc. UART port is used for the communication between SIM800L module and microcontroller. Level translation is built in with this module, so that microcontrollers of voltage up to 2.8 V can be used with it. The board has a Mobile positioning system which can be used to fetch the user location [6] (Figure 9.7). This module is used for sending the SMS to the registered mobile numbers.

9.3.5 CONNECTING SIM800L GSM MODULE TO ARDUINO UNO

First, insert the SIM card in the socket. Connect the pin-3 of Arduino board to the TX pin of the module as software serial is used for writing code to the module. Arduino's digital pin uses a 5 V GPIO but the SIM800L module has

3.3 V hence we cannot interface them directly so we have to step down to 3.3 Voltage so that the SIM800L module will not be damaged. A resistor can be used to achieve the voltage stepping down process. Around 10 k resistor would be fine between the sim module and Arduino d2, and a 20 k resistance would be fine between sim module and ground pin.

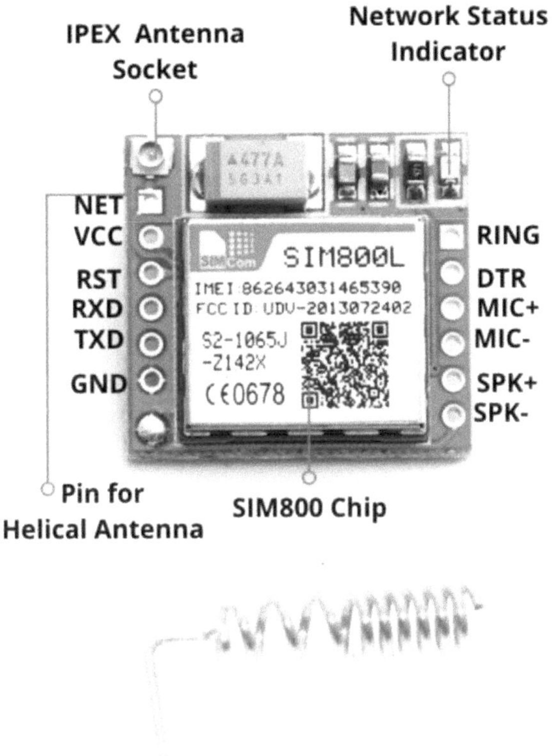

FIGURE 9.7 SIM800L module and antenna [22].

9.3.6 OPTICAL HEART-RATE SENSOR MODULE (FIGURE 9.8)

Optical sensors for heart rate monitor use infrared light source for measuring the heart rate. The sensor flashes the infrared light towards the human body and measures the light reflected with the help of photodiode.

Hemoglobin which is the primary component of blood has some characteristics to absorb the incident light, so by observing the flow of blood which changes on contraction and release of heart we can identify and measure the heart rate of the body and use the data for further processing [24] (Figure 9.9).

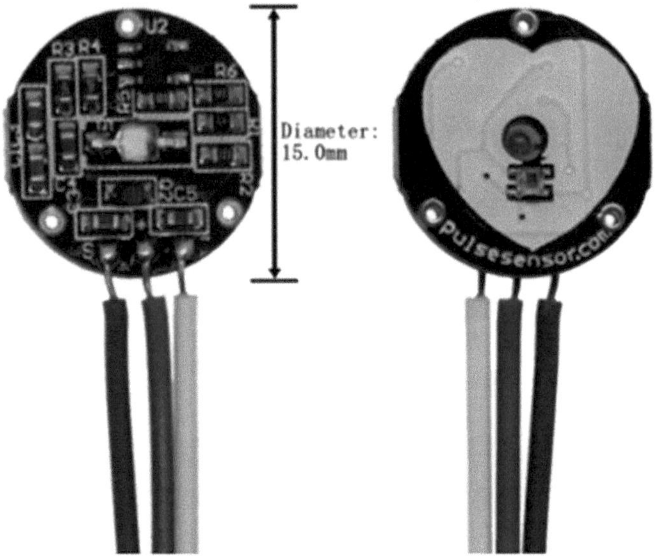

FIGURE 9.8 Heart-pulse sensor [23].

Note: Voltage supply between 5 V and 3.3 V; analog signal is received as output; output signal size: 0 ~ 3.3 V (3.3 V power supply) or 0 ~ 5 V (5 V power supply); current supply of 4 Mill Ampere; 16 mm diameter of circuit board with 1.6 mm thickness; and it has a maximum wavelength of 515 nm.

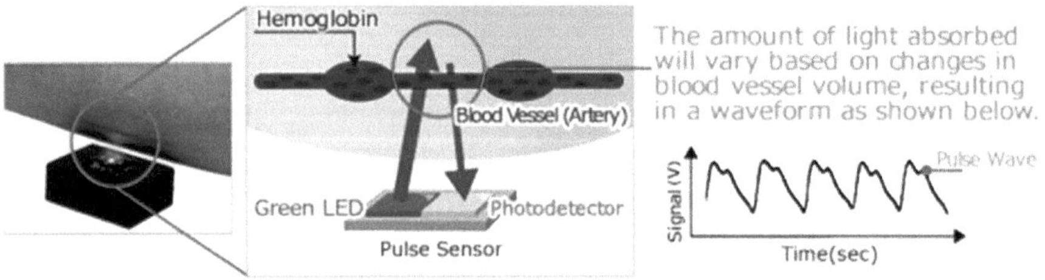

FIGURE 9.9 Working principle of heart-pulse sensor [24].

Since sunlight contains lot of infrared lights it affects the measuring of pulse using especially in outdoor which prevents stable operation of the sensor. This makes the device usable only in indoor and semi-indoor conditions [24].

For measuring the pulse wave in outdoor condition, a green light source can be used which has a similar absorption ration in hemoglobin and least affected by the sunlight thus this module uses a green LED as the light source [24].

9.4 WORKING PRINCIPLE OF THE DEVICE

The sensors are placed on the temples of the glasses, and both of them are connected to the Arduino board through cables. First the mobile numbers for which the SMS has to be sent will loaded into the program which is hardcoded for now but for the final product options will be provided to add the custom mobile numbers. 1x 9.0 V battery powers the Arduino board, 1x 9.0 V battery powers the thermostat temperature controller and 1x 9.0 V battery powers the buzzer. The buzzer and SIM800L modules are connected to the Arduino board.

Whenever the person wears the spectacles, the sensors placed in the temples of the glasses starts to monitor the blood pressure and the heart rate of the patient and constantly send the obtained data to the Arduino board where a python program will be running in the background and listening to the values. For each pair of values the python code performs a specific checking methodology, the python program takes the value of the blood pressure and checks if the value is in the range between 90/60 mmHg and 120/80 mmHg if yes then it does nothing but if the blood pressure value exceeds 120/80 mmHg or goes below 90/60 mmHg then it immediately gets the current temperature value and checks if it is in the range between 95°F and 100.4°F if it is so then the program just logs a warning message about the blood pressure and continues monitoring further values but if the temperature value is above 100.4°F or below 95°F then the program identifies this condition as a pre-symptom for seizure and immediately notifies the patient through the buzzer which indicates that they have high chances of getting a convulsion sooner and also sends the SMS to the registered mobile numbers about the condition of the patient with their location.

9.4.1 PROGRAM LOGIC FOR MANIPULATING THE VALUES FROM HEART PULSE

```
#include <Adafruit_GFX.h>
#include <Adafruit_SSD1306.h>
#include <SPI.h>
#include <Wire.h>
```

```
#define sr A0
#define setMax 540
Adafruit_SSD1306 con = Adafruit_SSD1306(128, 64, &Wire);
int valX = 0;
int valY = 60;
int senX = 0;
int SensorVal;
int value;
long TimeX = 0;
long TimeL = 0;
int count = 0;
int Bpm = 0;
void setup() {
Serial.begin(9600);
con.begin(SSD1306_SWITCHCAPVCC, 0x3C);
delay(1000);
con.clearDisplay();
}
void loop() {
SensorVal = analogRead(sr);
Serial.println(SensorVal);
value = map(SensorVal, 0, 1024, 0, 45);
int y = 60 – value;
If (x > 128) {
x = 0;
valX = 0;
con.clearDisplay();
}
con.drawLine(valX, valY, x, y, WHITE);
valX = x;
valY = y;
x ++;
findBP();
con.setCursor(0, 0);
con.setTextSize(2);
con.setTextColor(SSD1306_WHITE);
con.print("BP");
con.display();
}
```

```
void findBP() {
if (SensorVal > setMax) {
TimeX = millis() – TimeL;
count++;
if (TimeX/1000 >= 60) {
TimeL = millis();
Serial.println(count);
con.setCursor(60, 0);
con.setTextSize(2);
con.setTextColor(SSD1306_WHITE);
con.display();
count = 0;
} } }
```

9.4.2 PROGRAMMING LOGIC FOR SENDING SMS USING SIM 800L MODULE

```
#include <SoftwareSerial.h>
SoftwareSerial serial1(3, 2);
void setup()
{
Serial.begin(9600);
serial1.begin(9600);
Serial.println("Starting");
delay(1000);
serial1.println("AT");
update();
serial1.println("AT+CMGF=1");
update();
serial1.println("AT+CMGS=\"+XXnnnnnnnnnn\"");
update();
update();
serial1.write(26);
}
void loop()
{}
void update()
{
delay(350);
```

```
while (Serial.available())
{
serial1.write(Serial.read());
}
while(serial1.available()) {
Serial.write(serial1.read());
}
}
```

9.5 CONCLUSION

The ideology of this project is not just a seizure detecting device, the usage of IoT in the field of medical and healthcare to enhance them is the core idea of this project. There are lots of IoT enabled devices out there in market where most of them are designed to make the day-to-day tasks easier and making the humans even lazier, but the thing is IoT has a lot of potential which has not been well utilized yet. IoT can not only make the day-to-day task easier, but it can also do a lot more if it is used in the proper field for a proper way. In our project we propose that the IoT have the potential to be a lifesaver too in certain situations. This prototype model is just to demonstrate the proof of concept. The actual goal is to create the spectacles with all the functionalities resides inside the glasses itself without an external Arduino board. The spectacles will be connected to the user's Smartphone and all the data will be logged in a database which will make it really easy to doctors for analyzing the patient. The IoT is used widely in the field of automation and other such IT fields, but it will be also effective if we use it in the medical field.

KEYWORDS

- **Arduino board**
- **biosignals**
- **ideology**
- **internet of things**
- **seizure and symptoms**
- **temperature sensor**
- **thermostat temperature controller module**

REFERENCES

1. Gregg, N. M., & Crepeau, A. Z., (2021). Devices to monitor and track the patient's epilepsy. *Epilepsy*, 301–343.
2. Kramer, U., Kipervasser, S., Shlitner, A., & Kuzniecky, R., (2011). A novel portable seizure detection alarm system: Preliminary results. *Journal of Clinical Neurophysiology, 28*(1), 36–38.
3. Krishnu, R. S., Shukla, R., Chattar, H. K., Paul, B. S., Kaur, R., Khokhar, A., & Sahani, A. K., (2022). An emergency message and call system for people with epilepsy. In: *2022 IEEE International Instrumentation and Measurement Technology Conference (I2MTC)* (pp. 1–5). IEEE.
4. Hanke, J. M., Schindler, K. A., & Seiler, A., (2022). On the relationships between epilepsy, sleep, and Alzheimer's disease: A narrative review. *Epilepsy and Behavior, 129*, 108609.
5. Preetha, J., Sasikumar, G., Andrzej, S., Parameshachari, B. D., & Jewel, S., (2021). Collision-aware routing using multi-objective seagull optimization algorithm for WSN-based IoT. *Sensors, 21*(24), 8496. https://doi.org/10.3390/s21248496.
6. Preetha, J., Manirathnam, M., Chaitanya, A., & Prakash, R. R., (2020). Raspberry Pi-based face recognition system. *International Journal of Engineering Research and Technology (IJERT), 8*(8). ISSN: 2278-0181.
7. https://nettigo.eu/products/sim800l-gsm-grps-module#:~:text=SIM800L%20is%20a%20miniature%20cellular,that%20require%20long%20range%20connectivity (accessed on 20 November 2023).
8. Rahil, B., Said, M. B., & Boujemaa, H., (2020). Fuzzy C-means based clustering algorithm in WSNs for IoT applications. In: *Proceedings of the 2020 International Wireless Communications and Mobile Computing (IWCMC)* (pp. 126–130). Limassol, Cyprus.
9. https://www.rohm.com/sensor-shield-support/heart-rate-sensor (accessed on 20 November 2023).
10. Bahr, A., Matthias, S., Maria, A. F., Hendrik, M. L., Igor, B., Anna-Sophia, B., Peer, W., Thomas, S., & Franz, F., (2021). Epileptic seizure detection on an ultra-low-power embedded RISC-V processor using a convolutional neural network. *Biosensors, 11*(7), 203. https://doi.org/10.3390/bios11070203 (accessed on 30 November 2023).
11. https://www.electrokit.com/uploads/productfile/41015/Active_Piezo_Buzzer.pdf (accessed on 20 November 2023).
12. Ahmad, M., Shah, B., Ullah, A., Moreira, F., Alfandi, O., Ali, G., & Hameed, A., (2021). Optimal clustering in wireless sensor networks for the internet of things based on memetic algorithm: Meme WSN. *Wirel. Commun. Mob. Comput., 2021*, 8875950.
13. https://en.wikipedia.org/wiki/Internet_of_things (accessed on 20 November 2023).
14. Praveen, K. M., & Rajasekhara, B. M., (2019). A hybrid cluster head selection model for internet of things. *Clust. Comput., 22*, 13095–13107.
15. https://create.arduino.cc/projecthub/JANAK13/using-sensors-with-arduino-eab1ec; https://www.techtarget.com/iotagenda/definition/Internet-of-Things-IoT (accessed on 20 November 2023).
16. https://lastminuteengineers.com/sim800l-gsm-module-arduino-tutorial/#:~:text=Wiring%20%E2%80%93%20Connecting%20SIM800L%20GSM%20module%20to%20Arduino%20UNO&text=Start%20by%20soldering%2Fconnecting%20the,to%20talk%20to%20the%20module (accessed on 20 November 2023).

17. https://create.arduino.cc/projecthub/Ingeimaks/diy-heart-rate-sensor-a96e89 (accessed on 20 November 2023).
18. https://www.stjohnvic.com.au/news/managing-epilepsy-provide-first-aid/ (accessed on 20 November 2023).
19. https://www.mayoclinic.org/diseases-conditions/seizure/symptoms-causes/syc-20365711 (accessed on 20 November 2023).
20. https://quartzcomponents.com/products/digital-temperature-controller-thermostat-module-w1209 (accessed on 20 November 2023).
21. https://learn.sparkfun.com/tutorials/what-is-an-arduino/all (accessed on 20 November 2023).
22. https://1ohm.in/product/sim800l-gprs-gsm-module-with-wire-antenna/ (accessed on 20 November 2023).
23. https://www.aliexpress.com/item/601121205.html (accessed on 20 November 2023).
24. https://www.rohm.com/electronics-basics/sensor/pulse-sensor (accessed on 20 November 2023).

CHAPTER 10

A TOPSIS METHOD FOR INDUSTRIAL SERVICE ROBOT SELECTION WITH DIFFERENT WEIGHT APPROACH

D. DEEPA[1], S. SURESH[2], K. G. SARAVANAN[3], and RAGHAVENDRA DATTA NAIK[4]

[1]*Department of Mechanical Engineering, Muthayammal Engineering College, Rasipuram, Namakkal, Tamil Nadu, India*

[2]*Department of Mechanical Engineering, Erode Sengunthar Engineering College, Erode, Tamilnadu, India*

[3]*Department of Mechanical Engineering, Sona College of Technology, Salem, Tamil Nadu, India*

[4]*Department of Mechanical Engineering, Goa College of Engineering, Goa, India*

ABSTRACT

In recent years, one of the most difficult problems in the production environment, based on automation and smartness for real-time decision-making, is the selection of a robot for particular industrial applications. Currently, there are a number of industrial service robots on the market, each with its own set of capabilities, features, facilities, and specifications. In this chapter, technique for order of preference by similarity to ideal solution (TOPSIS) approach has been used to select the best-suited robot to attain the desired output with precise application ability, and minimum cost. Four objective criteria elicitation approaches, namely, equal weightage approach, entropy approach, AHP approach, and critic approach, are proposed and compared to indicate individual decision makers.

Robotics and Automation in Healthcare: Advanced Applications. R. Thanigaivelan, Sanjay Singh, & Clement Christy Deepak C. (Eds.)

In equal weightage method all the criteria were given equal importance. But in other methods various weightages were given for each criterion. Finally, CINCINNATI satisfied all the conditions, and it ranked top among other Robots.

10.1 INTRODUCTION

An industrial robot is commonly employed in manufacturing applications in industries. Robots are utilized in welding, painting, assembly and disassembly, packaging, and labeling, pick and place for printed circuit boards, palletizing, and product inspection. It decreases the workload imposed on a human. Each robot does have a high level of endurance, speed, and reliability [1, 2].

Control accuracy, precision, consistency, load carrying capacity, program versatility, maximum tip velocity, memory capacity, and supplier service quality are the most crucial factors to take into account when choosing an industrial robot for a particular application [3–5]. These factors influencing robot selection can be divided into objective and subjective factors, as well as beneficial and non-beneficial factors. Objective attributes, such as a robot's cost and weight carrying capacity, can be numerically defined. Subjective attributes, on the other hand, are qualitative in nature, such as the quality of a vendor's service, a robot's programming flexibility, and so on. Beneficial features have greater values that are always desirable, such as load bearing capacity and programming flexibility, while non-beneficial attributes have lower values that are preferable, such as cost and repeatability [6–8].

When choosing an industrial robot for a certain application, the decision maker must take into account all of performance metrics between the robot [9]. Because there are so many robots on the market, choosing the best one for a certain application can be difficult. Improper robot selection can have a severe impact on not just productivity and product quality, but also on the reputation of industrial companies. Executing a robot application, on the other hand, is a capital-intensive task. As a result, prior to its implementation, a thorough study of its feasibility and performance is required, as well as an assessment of the influence of various selection criteria [10]. Previous scholars have proposed applications of multi-criteria decision-making (MCDM) approaches [9], production system performance optimization models, extended gray relational analysis (GRA) method [12, 13], MTLBO algorithm [14], computer-assisted models, and statistical models, among others.

Goswami et al. [15] presented a robot selection process based on MCDM Models of TOPSIS-ARAS and COPRAS-ARAS, as well as a comparison of the relative ranks of various robots obtained using these two methods. Additionally, a coding system is employed to specify different aspects of robot selection, and a merit rating is applied to rate the robots according to how well they are suited for a given industrial application. Moreover, sensitivity analysis was used to evaluate the effects of weight change and to validate the robustness of the MCDM procedures that had been deployed. Chatterjee et al. [16] used compromise ranking and outranking methods for selecting industrial robots and reported that the relative ranks of the alternative robots acquired using these two MCDM approaches appear to match those obtained from previous researchers very well.

For the purpose of choosing robots, the analytical hierarchy process (AHP) method can concurrently take both subjective and objective criteria into account [17]. Khandekar et al. [18] applied fuzzy axiomatic design (FAD) principles to explore for solving a real time robot selection problem. It is reported that the FAD established the systematic and dependable tool to solve industrial robot selection problem.

Parkan & Wu [19] examined the performance of the OCRA and TOPSIS methodologies in a robot selection problem and described their applications and interrelationships. Both of these methods are found to be highly interconnected, with the same mathematical treatment applied to their performance assessments and decision-making processes, despite their apparent structural differences. To solve robot selection challenges, Khouja [20] used a two-phase paradigm. Data envelopment analysis (DEA) was utilized in the first phase to discover feasible robot alternatives based on satisfaction of some predetermined performance indicators, and an MCDM model was used in the second phase to pick the best robot among those identified in the first phase. Karsak et al. [11] proposed an industrial robot selection decision model based on fuzzy linear regression. The TOPSIS method (technique for order preference by resemblance to ideal solution) could be used to rank options and offer a solution to the decision maker. The chosen solution should be the smallest distance from both the optimum solutions, both positive and negative.

It is observed that the past researchers have adopted different variants of MCDM method to solve the robot selection problems. No attempt has yet been made to compare the different weightage models using TOPSIS methods. This chapter evaluates the performance of different industrial service robot considering different attributes using TOPSIS method.

10.2 METHOD

The proposed weightage methods are applied with the consideration of five attributes: Load capacity, repeatability, maximum tip speed, memory capacity, and manipulator reach:

- Load capacity (C1), i.e., the robot's maximum transportable weight;
- Repeatability (C2), i.e., the measure of the accuracy with which the robot permits the end effectors to return to a specific point;
- Maximum tip speed (C3), i.e., optimizing motion speed and acceleration according to the load on the tip of the robot arm;
- Memory capacity (C4), i.e., amount of memory space given to it;
- Manipulator reach (C5).

Seven robots from different companies like ASEA (A1), CINCINNATI (A2), CYBOTECH (A3), HITACHI (A4), PUMA (A5), US (A6), YASKAWA (A7) were taken for the analysis. In that maximum tip speed, Load capacity, manipulator reach and memory capacity, was maintained high and repeatability was maintained low. The decision matrix with seven alternative robots and six attributes is given in Table 10.1.

TABLE 10.1 Quantitative Data for Robot Selection

SL. No.	Types of Robots	Load Capacity (kg) (C_1)	Repeatability (C_2)	Max. Tip Speed (mm/s) (C_3)	Memory Capacity (C_4)	Manipulator Reach (mm) (C_5)
1.	A_1	60	0.40	2,540	500	990
2.	A_2	6.35	0.15	1,016	3,000	1,041
3.	A_3	6.8	0.10	1727.2	1,500	1,676
4.	A_4	10	0.20	1,000	2,000	965
5.	A_5	2.5	0.10	560	500	915
6.	A_6	4.5	0.08	1,016	350	508
7.	A_7	3	0.10	177	1,000	920

10.3 MULTI-CRITERIA DECISION-MAKING (MCDM) METHOD

10.3.1 TOPSIS METHODOLOGY

TOPSIS is a multi-criteria decision-making tool that is used to select a best solution among large number of alternatives. The alternative with the highest distance from the negative ideal solution (NIS) and the shortest difference from the positive ideal solution (PIS) will be chosen. In TOPSIS, the response parameter to be maximized is considered as beneficial attribute and response to be minimized is considered as non-beneficial attribute. The steps followed in TOPSIS are:

- **Step 1:** The decision matrix is to be constructed with response data. The decision matrix should contain n number of attributes and m number of alternatives solutions. Here attribute denotes the various responses and alternative solution denotes the experimental trial.

$$D = \begin{bmatrix} a_{11} & a_{12} & \cdots & a_{1n} \\ a_{21} & a_{21} & \cdots & a_{21} \\ \vdots & \vdots & \vdots & \vdots \\ a_{m1} & a_{m1} & \cdots & a_{mn} \end{bmatrix}$$

- **Step 2:** Each attribute own data in different ranges, hence normalization is to be done to convert the range of data between 0 and 1 for easy calculation.

$$N_{ij} = \frac{a_{ij}}{\sqrt{\sum_{i=1}^{m} a_{ij}^2}} \tag{1}$$

where; N_{ij} is the normalized value for i = 1, 2, ..., m and j = 1, 2, ..., n.

- **Step 3:** Weighted normalized matrix is calculated by assigning appropriate weights to each attribute. Always, the sum of weights should be equal to 1.

$$\delta ij = W_j \times N_{ij} \tag{2}$$

where; $\sum_{j=1}^{n} W_j = 1$

- **Step 4:** Determination of PIS and NIS from weight normalized matrix.

$$\delta^+ = \left(\delta_1^+, \delta_2^+, \ldots\ldots \delta_n^+\right) = \begin{cases} \left(\max \delta_{ij} \,|\, j \in J_1\right) \\ \left(\min \delta_{ij} \,|\, j \in J_2, i = 1, 2, \ldots\right) \end{cases} \tag{3}$$

$$\delta^- = \left(\delta_1^-, \delta_2^-, \ldots\ldots \delta_n^-\right) = \begin{cases} \left(\min \delta_{ij} \,|\, j \in J_1\right) \\ \left(\max \delta_{ij} \,|\, j \in J_2,\right) \end{cases} \tag{4}$$

where; J_1 belongs to beneficial attribute; J_2 belongs to non-beneficial attribute; and i = 1, 2, ..., n.

- **Step 5:** Calculating the Euclidian distance of each alternative from ideal solutions.

$$S_i^+ = \sqrt{\sum_{j=1}^{1}\left(\delta_{ij} - \delta^+\right)^2} \quad (5)$$

$$S_i^- = \sqrt{\sum_{j=1}^{1}\left(\delta_{ij} - \delta^-\right)^2} \quad (6)$$

where; $i = 1, 2, \ldots, m$.

- **Step 6:** the closeness coefficient of each alternative solution is calculated using the following equation:

$$CC = \frac{S_i^-}{S_i^+ - S_i^-} \quad (7)$$

- **Step 7:** Rank the results in descending order according to the values of CC in step 6.

10.3.2 EQUAL WEIGHTAGE APPROACH

The equal weightage as mentioned in Table 10.2 is applied to all the attributes. The Quantitative dates under different attributes are normalized using the formula 1 and tabulated in Table 10.3. PIS, Negative ideal solution, and ranking are then calculated for all the criteria, as shown in Tables 10.4 and 10.5 using Formula 5 and 6, respectively. The results and final ranking of robots under equal weightage approach are shown in Table 10.6.

TABLE 10.2 Weightage for Attributes

Attributes	C_1	C_2	C_3	C_4	C_5
Weightage	0.2	0.2	0.2	0.2	0.2

TABLE 10.3 Normalized Decision Matrix

Types of Robots	C_1	C_2	C_3	C_4	C_5
A_1	0.194102	0.157232	360.03	12.172	70.43
A_2	0.02054	8.844×10^{-3}	57.60	438.21	77.875
A_3	0.02199	3.93×10^{-3}	166.47	109.55	201.85
A_4	0.03235	0.01572	55.805	194.760	66.919
A_5	8.0876×10^{-3}	3.93×10^{-3}	17.500	12.1724	60.164
A_6	0.01455	2.514×10^{-3}	57.605	5.9644	18.5450
A_7	9.70×10^{-3}	3.93×10^{-3}	1.74832	48.69	60.824

TABLE 10.4 Positive-Ideal Solution

Types of Robots	C_1	C_2	C_3	C_4	C_5
A_1	0	0.023937	0	181508.37	17271.2164
A_2	0.030123	0.00633	91463.90	0	15369.800
A_3	0.029622	2.005056×10^{-6}	37465.47	108017.39	0
A_4	0.026163	0.00017439	92552.850	59267.90	18206.374
A_5	0.034601	2.005056×10^{-6}	117326.800	181508.03	20074.922
A_6	0.032238	0	94460.88	186836.25	33600.723
A_7	0.0340026	2.005056×10^{-6}	128365.76	151725.83	19888.3326

TABLE 10.5 Negative-Ideal Solution

Types of Robots	C_1	C_2	C_3	C_4	C_5
A_1	0.03460	0	128365.76	38.534	2692.053
A_2	1.55062×10^{-4}	0.022018	3119.410	186836.25	3520.048
A_3	1.93276	0.02350	27133.23	10729.976	33600.72
A_4	3.2672×10^{-4}	1.024×10^{-9}	2922.1246	35643.77	2340.043
A_5	7.02960×10^{-4}	0.023501	248.115	38.539	1732.1411
A_6	6.4624×10^{-3}	0.0239376	3119.968	0	0
A_7	2.6127×10^{-6}	0.02350	0	1825.476	1787.513

TABLE 10.6 Results of Relative Closeness and Ranking

Types of Robots	$S_i^1/S_i^* + S_i^1$	Rank
A_1	0.44815	2
A_2	0.57369	1
A_3	0.389816	3
A_4	0.32908	4
A_5	0.073699	7
A_6	0.09092	6
A_7	0.098892	5

10.3.3 ENTROPY METHOD

Entropy of the uncertainty in data that probability theory expresses. The steps that make up this method's process are described below:

- **Step 1:** The performance matrix is first normalized using the formulae indicated in Table 10.7 for benefit and cost attributes.

$$r_{ij} = \frac{a_{ij}}{\sum_{i=1}^{m} a_{ij}} \quad for\ i = 1, 2, \ldots, m \tag{8}$$

- **Step 2:** The following equation is used to calculate the values of entropy. The calculated r_{ij} ln r_{ij} value and obtained Entropy values of all the attributes are shown in Tables 10.8 and 10.9, respectively.

$$e_j = -(\ln m)^{-1} \sum_{i=1}^{m} r_{ij} \ln r_{ij} \quad for\ j = 1,2,\ldots,n \tag{9}$$

- **Step 3:** The following equation is used to calculate the attribute weights, which are then listed in Table 10.10.

$$W_j = \frac{1-e_j}{(n-\sum_{j=1}^{n} e_j)} \quad for\ j = 1,2,\ldots,n \tag{10}$$

Normalized values, PIS, Negative ideal solution, and ranking are then calculated for all the criteria, as shown in Tables 10.11–10.13, respectively using Eqns. (1), (5), and (6). The results and final ranking of robots are shown in Table 10.14.

TABLE 10.7 Finding r_{ij}

Types of Robots	Load Capacity (kg) (C_1)	Repeatability (C_2)	Max Tip Speed (C_3)	Memory Capacity (C_4)	Manipulator Reach (C_5)
A_1	0.6441	0.3539	0.3160	0.05649	0.141126
A_2	0.06816	0.1327	0.1264	0.33898	0.14839
A_3	0.07300	0.0884	0.2144	0.16949	0.2389
A_4	0.10735	0.17699	0.1244	0.22598	0.13756
A_5	0.0268	0.0884	0.06968	0.05649	0.13043
A_6	0.048309	0.0707	0.1264	0.03954	0.07241
A_7	0.03220	0.0884	0.0220	0.11299	0.131147

TABLE 10.8 Finding of r_{ij} ln r_{ij}

Types of Robots	Load Capacity (kg) (C_1)	Repeatability (C_2)	Max Tip Speed (C_3)	Memory Capacity (C_4)	Manipulator Reach (C_5)
A_1	–0.12305	–0.15965	–0.15809	–0.07050	–0.120012
A_2	–0.07950	–0.116395	–0.11353	–0.15926	–0.12295
A_3	–0.08297	–0.093133	–0.14350	–0.13065	–0.14854
A_4	–0.10404	–0.133105	–0.11260	–0.14596	–0.11850
A_5	–0.04212	–0.09313	–0.08061	–0.07050	–0.11538
A_6	–0.06357	–0.08134	–0.1135	–0.05547	–0.08256
A_7	–0.04804	–0.09313	–0.03646	–0.10699	–0.11570

TABLE 10.9 Entropy Values

Attributes	C_1	C_2	C_3	C_4	C_5
E_{ij}	0.27919	0.04785	0.018736	0.05498	0.05945

TABLE 10.10 Weightage for Attributes by Entropy Approach

Attributes	C_1	C_2	C_3	C_4	C_5
Weightage	0.159	0.210	0.216	0.208	0.207

TABLE 10.11 Normalized Value

Types of Robots	Load Capacity (kg) (C_1)	Repeatability (C_2)	Max Tip Speed (C_3)	Memory Capacity (C_4)	Manipulator Reach (C_5)
A_1	0.1543	0.1650	388.836	12.659	72.897
A_2	0.01632	9.286×10^{-3}	62.2138	455.738	80.6012
A_3	0.017488	4.1265×10^{-3}	179.798	113.934	208.923
A_4	0.02571	0.01650	60.2698	202.550	69.26199
A_5	6.429×10^{-3}	4.126×10^{-3}	18.900	12.659	62.270
A_6	0.01157	2.639×10^{-3}	62.2138	6.2029	19.1941
A_7	7.714×10^{-3}	4.126×10^{-3}	1.888	50.6376	62.9530

TABLE 10.12 Positive Ideal Solution

Types of Robots	Load Capacity (kg) (C_1)	Repeatability (C_2)	Max Tip Speed (C_3)	Memory Capacity (C_4)	Manipulator Reach (C_5)
A_1	0	0.02636	0	196319	18503.07
A_2	0.01903	4.418×10^{-5}	106682.06	0	16466.48
A_3	0.01871	2.2126×10^{-6}	43696.88	116829.97	0
A_4	0.01653	1.9212×10^{-4}	107955.74	64104.16	19505.1977
A_5	0.02186	2.211×10^{-6}	136852.64	196319	21507.1024
A_6	0.02037	0	106682.06	202081.80	35997.055
A_7	0.02148	2.211×10^{-6}	149728.75	164106.33	21307.2409

TABLE 10.13 Negative Ideal Solution

Types of Robots	Load Capacity (kg) (C_1)	Repeatability (C_2)	Max Tip Speed (C_3)	Memory Capacity (C_4)	Manipulator Reach (C_5)
A_1	0.02186	0	149728.75	41.6812	2884.00
A_2	9.801×10^{-5}	0.02424	3639.20	202081.80	3770.831
A_3	1.225×10^{-5}	0.025880	31651.96	11605.98	35997.055
A_4	3.721×10^{-4}	0	3408.4345	38552.18	2506.793
A_5	0	0.025880	289.408	41.6812	1855.533
A_6	2.6522×10^{-6}	0.02636	3639.202	0	0
A_7	1.6744×10^{-6}	0.025880	0	1974.442	1914.841

TABLE 10.14 Results of Relative Closeness and Ranking

Types of Robots	$S_i^1/S_i^* + S_i^1$	Rank
A_1	0.4573	2
A_2	0.5660	1
A_3	0.41267	3
A_4	0.32514	4
A_5	0.07280	7
A_6	0.093169	6
A_7	0.09724	5

10.3.4 ANALYTICAL HIERARCHY PROCESS (AHP) WEIGHTING METHOD

The detailed analytic process of the AHP method is as follows:

- Establish the comparison matrix C according to the decision-maker's opinion. The Pair-wise comparison Matrix of all attributes of selected robots is shown in Table 10.15.
- Calculate the maximum eigenvalues λ_{max} and eigenvectors w of matrix C, and the weight vector, of each ωj evaluation index is obtained after normalization. The calculated maximum eigenvalues are listed in Table 10.16.
- Consistency test with ratio CR.

$$CR = CI/RI$$

The attribute weights are determined and listed in Table 10.17.

Normalized values, PIS and Negative ideal solution are then calculated for all the criteria, as shown in Tables 10.18–10.20, respectively using Formula 1, 5, and 6. The results and final ranking of robots are shown in Table 10.21.

TABLE 10.15 Pair-Wise Comparison Matrix

	Load Capacity (C_1)	Repeatability (C_2)	Max Tip Speed (C_3)	Memory Capacity (C_4)	Manipulator Reach (C_5)
Load capacity (C_1)	1	3	5	1/3	1/3
Repeatability (C_2)	1/3	1	1/3	1/5	1/3
Max tip speed (C_3)	1/5	3	1	5	3
Memory capacity (C_4)	3	5	1/5	1	3
Manipulator reach (C_5)	3	3	1/3	1/3	1
$\sum_{Sum}$	7.5333	15	6.8666	6.8666	7.6666

TABLE 10.16 Calculation of Maximum Eigen Values

	Load Capacity (C_1)	Repeatability (C_2)	Max Tip Speed (C_3)	Memory Capacity (C_4)	Manipulator Reach (C_5)
Load capacity (C_1)	0.1327	0.2	0.7281	0.0485	0.0435
Repeatability (C_2)	0.0442	0.0666	0.04853	0.02913	0.0435
Max tip speed (C_3)	0.0265	0.2	0.14563	0.7282	0.39130
Memory capacity (C_4)	0.3982	0.3333	0.029126	0.1456	0.39130
Manipulator reach (C_5)	0.3982	0.2	0.04854	0.0485	0.13044

TABLE 10.17 Weightage for Attributes by AHP Approach

Attributes	C_1	C_2	C_3	C_4	C_5
Weightage	0.23056	0.0464	0.2983	0.2719	0.1651

TABLE 10.18 Normalized Vales

Types of Robots	Load Capacity (kg) (C_1)	Repeatability (C_2)	Max Tip Speed (C_3)	Memory Capacity (C_4)	Manipulator Reach (C_5)
A_1	0.97051	0.78616	1800.171	60.862	352.160
A_2	0.1027	0.04422	288.027	2191.05	389.37
A_3	0.10999	0.01965	832.399	547.763	1009.29
A_4	0.16175	0.0786	279.027	973.801	334.599
A_5	0.040438	0.01965	87.502	60.862	300.824
A_6	0.07278	0.01257	288.027	29.822	92.7252
A_7	0.04852	0.01965	8.7416	243.450	304.1208

TABLE 10.19 Positive Ideal Solution

Types of Robots	Load Capacity (kg) (C_1)	Repeatability (C_2)	Max Tip Speed (C_3)	Memory Capacity (C_4)	Manipulator Reach (C_5)
A_1	0	1.288×10^{-4}	0	335470.32	11770.73
A_2	0.04004	2.1567×10^{-6}	203466.85	0	10475.31
A_3	0.04782	1.0795×10^{-7}	83340.18	199639.17	0
A_4	0.03479	9.386×10^{-6}	205896.32	270274.17	12407.95
A_5	0.046000	1.0795×10^{-7}	261007.77	335470.32	13681.27
A_6	0.04285	0	203466.49	345317.71	22899.25
A_7	0.04520	1.0795×10^{-7}	285565.61	280425.320	13554.268

TABLE 10.20 Negative Ideal Solution

Types of Robots	Load Capacity (kg) (C_1)	Repeatability (C_2)	Max Tip Speed (C_3)	Memory Capacity (C_4)	Manipulator Reach (C_5)
A_1	0.0478	0	285565.61	71.223	1834.580
A_2	3.4510×10^{-4}	1.1853×10^{-3}	6940.622	345317.71	2398.64
A_3	0	1.26509×10^{-3}	60366.72	19832.35	22899.25
A_4	1.036×10^{-3}	0.03283	6500.455	4591.200	1594.724
A_5	1.780×10^{-5}	1.26509×10^{-3}	551.977	71.2234	1180.472
A_6	1.3635×10^{-4}	1.2885×10^{-3}	6940.689	0	0
A_7	3.705×10^{-5}	1.26509×10^{-3}	0	3373.913	1218.163

TABLE 10.21 Results of Relative Closeness and Ranking

Types of Robots	$S_i^1/S_i^* + S_i^1$	Rank
A_1	0.4529	2
A_2	0.6237	1
A_3	0.2670	3
A_4	0.02530	4
A_5	0.00294	7
A_6	0.01199	6
A_7	0.007861	5

10.3.5 CRITIC APPROACH

The following is a summary of the CRITIC approach's working method:

i. Normalizing the input data as listed in Table 10.22.
ii. Calculating the standard deviation for the linked criterion j, s, which measures the degree of contrast. In this sense, the value of criterion j to the decision-making process might be measured by s_{jj}.
iii. The next step is to calculate the index (C) by using Eqn. (6).

$$I_j = \sigma_j \bullet \sum_{k=1}^{n} (1 - r_{jk}) \tag{11}$$

Calculated index value of attributes is listed in Table 10.23.

iv. Calculating the weights of the criteria in relation to each criterion.

$$w_j = \frac{I_j}{\sum_{j=1}^{n} I_j}, \; j = 1, 2, \ldots, n. \tag{12}$$

The large the value I_j, the criterion's proportional value for the decision-making process increases with the amount of information that it reflects. The attribute weights are determined and listed in Table 10.24.

Normalized values, PIS and negative ideal solution are then calculated for all the criteria, as shown in Tables 10.25–10.27, respectively using Formula 1, 5, and 6. The results and final ranking of robots are shown in Table 10.28.

TABLE 10.22 Normalized Input Data

Types of Robots	Load Capacity (kg) (C_1)	Repeatability (C_2)	Max Tip Speed (C_3)	Memory Capacity (C_4)	Manipulator Reach (C_5)
A_1	1	0	0	0.056	0.412
A_2	0.066	0.78	0.355	1	0.456
A_3	0.07	0.937	0.655	0.433	1
A_4	0.1304	0.625	0.348	0.622	0.3912
A_5	0	0.937	0.162	0.0566	0.348
A_6	0.034	1	0.355	0	0
A_7	0.008	0.937	0	0.2452	0.3527

TABLE 10.23 Conflict Values

Attributes	Load Capacity (kg) (C_1)	Repeatability (C_2)	Max Tip Speed (C_3)	Memory Capacity (C_4)	Manipulator Reach (C_5)
Load capacity (kg) (C_1)	0	1.961	1.44	1.270	0.980
Repeatability (C_2)	1.96	0	0.5586	0.923	1.0188
Max tip speed (C_3)	1.44	0.063	0	0.576	0.4766
Memory capacity (C_4)	1.270	0.923	0.5758	0	0.6296
Manipulator reach (C_5)	0.9802	0.063	0.4766	0.9434	0

TABLE 10.24 Weightage for Attributes by CRITIC Approach

Attributes	C_1	C_2	C_3	C_4	C_5
Weightage	0.3299	0.2542	0.0963	0.20171	0.1177

TABLE 10.25 Normalized Values

Types of Robots	Load Capacity (kg) (C_1)	Repeatability (C_2)	Max Tip Speed (C_3)	Memory Capacity (C_4)	Manipulator Reach (C_5)
A_1	0.97051	0.78616	1800.171	60.862	352.160
A_2	0.1027	0.04422	288.027	2191.05	389.378
A_3	0.10999	0.01965	832.399	547.763	1009.29
A_4	0.16175	0.0786	279.027	973.801	334.599
A_5	0.040438	0.01965	87.502	60.862	300.824
A_6	0.07278	0.01257	288.027	29.822	92.7252
A_7	0.04852	0.01965	8.7416	243.450	304.1208

TABLE 10.26 Positive Ideal Solution

Types of Robots	Load Capacity (kg) (C_1)	Repeatability (C_2)	Max Tip Speed (C_3)	Memory Capacity (C_4)	Manipulator Reach (C_5)
A_1	0	0.03865	0	184619.74	5983.02
A_2	1×10^{-4}	6.416×10^{-5}	21205.18	0	5324.62
A_3	0.0805	3.24×10^{-6}	8684.37	109872.36	0
A_4	0.0711	2.792×10^{-4}	21456.39	60284.9	6305.94
A_5	0.0940	2.92×10^{-6}	27199.92	184619.74	6952.89
A_6	0.0876	0	21205.18	190039.32	11637.447
A_7	0.0924	2.92×10^{-6}	29759.12	154331.12	6888.17

TABLE 10.27 Negative Ideal Solution

Types of Robots	Load Capacity (kg) (C_1)	Repeatability (C_2)	Max Tip Speed (C_3)	Memory Capacity (C_4)	Manipulator Reach (C_5)
A_1	0.0940	0	29759.12	6.261	931.89
A_2	3.880×10^{-4}	0.0355	194576.56	19039.32	1218.49
A_3	5.244×10^{-4}	0.0379	6291.39	10912.93	11637.44
A_4	1.6×10^{-3}	0.03236	677.47	36254.06	810.370
A_5	0	0.03798	57.522	39.200	599.90
A_6	1.144×10^{-4}	0.0386	722.98	0	0
A_7	7.29×10^{-6}	0.03798	0	1856.317	619.113

TABLE 10.28 Results of Relative Closeness and Ranking

Types of Robots	$S_i^1/S_i^* + S_i^1$	Rank
A_1	0.2863	4
A_2	0.7922	1
A_3	0.330	3
A_4	0.39566	2
A_5	0.0534	7
A_6	0.0538	6
A_7	0.1022	5

Table 10.29 exhibits the comparison of the rankings of the alternative robots as achieved using the different weightage approach Different methods (Equal weightage, Entropy method, AHP method and Critic method) and TOPSIS method. To summarize, the results indicated that three weightage methods out of four methods provided the same ranking.

TABLE 10.29 Comparison of Rankings

Equal Weightage	Entropy Method	AHP Method	Critic Method	Rank (Final)
2	2	2	4	2
1	1	1	1	1
3	3	3	3	3
4	4	4	2	4
7	7	7	7	7
6	6	6	6	6
5	5	5	5	5

10.4 CONCLUSIONS

In this work seven different robots such as A1, A2, A3, A4, A5, A6, and A7 are ranked on the basis of six criteria namely Load capacity, repeatability, maximum tip speed, Memory capacity and Manipulator reach, respectively. TOPSIS method is preferred to prioritize the criteria by providing different weightage approach (Equal weightage, Entropy method, AHP method and Critic method). The results of different weightage approach were integrated through the TOPSIS method to reveal the final ranking for industrial robot selection. According to the results obtained, A2 was the best alternative.

Among the seven robots, the selection order of robots is A2 > A1 > A3 > A4 > A7 > A6 > A5.

KEYWORDS

- **analytical hierarchy process**
- **data envelopment analysis**
- **fuzzy axiomatic design**
- **gray relational analysis**
- **multi-criteria decision-making**
- **robot**

REFERENCES

1. Xiong, G., Ding, Y., & Zhu, L., (2019). Stiffness-based pose optimization of an industrial robot for five-axis milling. *Robot. Comput. Integr. Manuf., 55*, 19–28. https://doi.org/https://doi.org/10.1016/j.rcim.2018.07.001.
2. Razif, M. R. M., Faudzi, A., Athif, M., Bavandi, M., Nordin, I. N. A. M., Natarajan, E., & Yaakob, O., (2014). Two chambers soft actuator realizing robotic gymnotiform swimmers' fin. In: *2014 IEEE International Conference on Robotics and Biomimetics (ROBIO 2014)* (pp. 15–20). IEEE. https://doi.org/10.1109/ROBIO.2014.7090300.
3. Luo, R. C., & Wang, H., (2018). Automated tool coordinate calibration system of an industrial robot. In: *2018 IEEE/RSJ International Conference on Intelligent Robots and Systems (IROS)* (pp. 5592–5597). IEEE. https://doi.org/10.1109/IROS.2018.8594298.
4. Story, M., Jaksic, C., Fletcher, S. R., Webb, P., Tang, G., & Carberry, J., (2021). Evaluating the use of human aware navigation in industrial robot arms. *Paladyn, J. Behav. Robot., 12*, 379–391. https://doi.org/doi:10.1515/pjbr-2021-0024.
5. Liu, W., Liu, Z., & Liu, S., (2013). Asymmetric cryptosystem using random binary phase modulation based on mixture retrieval type of Yang-Gu algorithm: Reply. *Opt. Lett., 38*, 4045. https://doi.org/10.1364/OL.38.004045.
6. Shiakolas, P. S., Conrad, K. L., & Yih, T. C., (2002). On the accuracy, repeatability, and degree of influence of kinematics parameters for industrial robots. *Int. J. Model. Simul., 22*, 245–254. https://doi.org/10.1080/02286203.2002.11442246.
7. Garcia, R. R., Bittencourt, A. C., & Villani, E., (2018). Relevant factors for the energy consumption of industrial robots. *J. Brazilian Soc. Mech. Sci. Eng., 40*, 464. https://doi.org/10.1007/s40430-018-1376-1.
8. Almurib, H., Al-Qrimli, H., & Thulasiraman, N., (2012). A review of application industrial robotic design. *Int. Conf. ICT Knowl. Eng.* https://doi.org/10.1109/ICTKE.2012.6152387.

9. Ali, A., & Rashid, T., (2021). Best–worst method for robot selection. *Soft Comput., 25*, 563–583. https://doi.org/10.1007/s00500-020-05169-z.
10. Mondal, S., & Chakraborty, S., (2013). A solution to robot selection problems using data envelopment analysis. *Int. J. Ind. Eng. Comput., 4*, 355–372. https://doi.org/10.5267/j.ijiec.2013.03.007.
11. Karsak, E. E., Sener, Z., & Dursun, M., (2012). Robot selection using a fuzzy regression-based decision-making approach. *Int. J. Prod. Res., 50*, 6826–6834. https://doi.org/10.1080/00207543.2011.627886.
12. Suresh, S., Venkatesan, K., Natarajan, E., & Rajesh, S., (2021). Performance analysis of nano-silicon carbide reinforced swept friction stir spot weld joint in AA6061-T6 alloy. *Silicon, 13*, 3399–3412. https://doi.org/10.1007/s12633-020-00751-4.
13. Suresh, S., Venkatesan, K., & Rajesh, S., (2019). Optimization of process parameters for friction stir spot welding of AA6061/Al_2O_3 by Taguchi method. In: *AIP Conference Proceedings* (pp. 1–10). https://doi.org/10.1063/1.5117961.
14. Suresh, S., Elango, N., Venkatesan, K., Lim, W. H., Palanikumar, K., & Rajesh, S., (2020). Sustainable friction stir spot welding of 6061-T6 aluminum alloy using improved non-dominated sorting teaching learning algorithm. *J. Mater. Res. Technol., 9*, 11650–11674 https://doi.org/https://doi.org/10.1016/j.jmrt.2020.08.043.
15. Goswami, S. S., Behera, D. K., Afzal, A., Razak, K. A., Khan, S. A., Rajendran, P., Subbiah, R., & Asif, M., (2021). Analysis of a robot selection problem using two newly developed hybrid MCDM models of TOPSIS-ARAS and COPRAS-ARAS. *Symmetry (Basel), 13*. https://doi.org/10.3390/sym13081331.
16. Chatterjee, P., Athawale, V. M., & Chakraborty, S., (2010). Selection of industrial robots using compromise ranking and outranking methods. *Robot. Comput. Integr. Manuf., 26*, 483–489. https://doi.org/10.1016/j.rcim.2010.03.007.
17. Goh, C. H., (1997). Analytic hierarchy process for robot selection. *J. Manuf. Syst., 16*, 381–386. https://doi.org/https://doi.org/10.1016/S0278-6125(97)88467-1.
18. Khandekar, A. V., & Shankar, C., (2015). Selection of industrial robot using axiomatic design principles in fuzzy environment. *Decis. Sci. Lett., 4*, 181–192. https://doi.org/10.5267/j.dsl.2014.12.004.
19. Parkan, C., & Wu, M. L., (1999). Decision-making and performance measurement models with applications to robot selection. *Comput. Ind. Eng., 36*, 503–523. https://doi.org/https://doi.org/10.1016/S0360-8352(99)00146-1.
20. Khouja, M., (1995). The use of data envelopment analysis for technology selection. *Comput. Ind. Eng., 28*, 123–132. https://doi.org/https://doi.org/10.1016/0360-8352(94)00032-I.

CHAPTER 11

AERIAL ROBOTICS: AN OVERVIEW OF THE MEDICAL APPLICATIONS OF DRONE SYSTEMS

C. RAMPRASADH

School of Mechanical Engineering, SASTRA Deemed to be University, Tirumalaisamudram, Thanjavur, Tamil Nadu, India

ABSTRACT

Drones are aerial vehicles possessing significant potential for utilization and application in improving people's quality of life. The initial sections of this chapter address their practical limits, including issues with GPS measurement latency and sensor attributes. Just as challenging as it is to fly a drone for any other scientific use, creating a drone specifically for medical usage is a challenging undertaking.

11.1 INTRODUCTION

Unmanned aerial systems, termed as drones, have become an increasingly important part of our daily lives compared to the preceding decade. Drones, otherwise called the Unmanned Aerial Vehicles (UAVs), have shown tremendous growth in their applications. In the past eight decades since their introduction in 1922, their applications have become widespread from civil to military to supply chain to healthcare systems. They are used in modern warfare as well. To our surprise, the first combat use of UAVs by the US Air Force was carried out in the year 1944 against Japan [1]. Today UAVs

Robotics and Automation in Healthcare: Advanced Applications. R. Thanigaivelan, Sanjay Singh, & Clement Christy Deepak C. (Eds.)

have grown as a frontier area of research and development depending upon their variety of usage domains which include intelligence, surveillance, target acquisition and reconnaissance (ISTAR), combat, vertical take-off, and landing (VTOL), radar and communication relay and aerial delivery and resupply. This chapter attempts to shed light on the drone systems and their applications in the medical and healthcare sector. But it would be better to start with a few fundamental theories of drone systems which would enable the reader to understand both the potentials and complexities in operating aerial systems.

11.2 CLASSIFICATION OF DRONES

UAVs are classified primarily based on their mission aspects and performance specifications. The performance specifications of a UAV are weight, cargo, range, speed, wing loading, price, engine kind, and power. When we look at an important factor based on which UAV is categorized is the weight. UAVs cover a wide range of weights from few grams to nearly 11 tons. According to their weight, UAVs are classified into three broad classes [1]. They are:

- **Class I:** Whose weight is less than 150 kg.
- **Class II:** Weighs between 150 and 600 kg.
- **Class III:** Vehicles are those which weigh above 600 kg.

Worldwide, there are 171 active types of UAVs in use. Among them, 107 are of Class I, 36 belong to Class II and remaining 28 are Class III type [1]. Figure 11.1 reproduced from Ref. [1] shows various drones of different classes giving a broad idea about the different types of drones in current world of applications. It clearly shows fixed wing vehicles, multirotors, and also helicopters. It is worth mentioning that the modern-day drone systems include flapping wing vehicles as well.

Though drones are widely used in a range of civil and military missions, there are advantages as well as certain challenges involved in operating them. At the same time, to their advantage, drones are compact and are capable of rapid deployment and real-time data acquisition. The major challenges involved in the operation of drone systems include weight constraints, aerodynamic behavior, nonlinear lift curves, atmospheric conditions, wind turbulence effects, sensor suites available, sensor characteristics, signal conditioning, and computational power required. Many mission aspects like hovering, reconnaissance, search and rescue, surveillance, etc., can be achieved. Some of the missions may include highly unconventional aspects such as flying the UAV with spatial constraints and also in the high Angle of Attack (AOA) region. Achieving hovering capability, which in turn enables vertical take offs and landing abolishing the requirement of runways and large spaces are also on the cards as potential research problems.

FIGURE 11.1 Examples of unmanned aerial vehicles according to classification.
Source: Adapted from Ref. [1]

But there are significant technical barriers which should be overcome before realizing unmanned flyers for any particular application including healthcare. This chapter in its later sections, elaborates such important barriers and the current trends focusing on medical applications. These include the low scale power generation and storage, navigation, and control, deployment of tiny sensors and the aerodynamics. One of the most intriguing and unclear elements of small-scale flight is their aerodynamics [2]. Researchers are continuously attempting to fly MAVs in highly unconventional missions like flying through windows and highly constrained space. These requirements may prove very essential in healthcare operations during emergency situations.

11.3 OPERATIONAL CHALLENGES

The main restriction on MAV comes in the form of their limited cubage and mass. Their flight is mainly disturbed by its interior and exterior factors. With smaller airplanes, all the aerodynamic rules become challenging. The most general problem associated with the aerodynamics of mini drones is that they are dominated by flows with very low Reynolds numbers. The design of efficient, compact vehicles poses a substantial aerodynamic challenge due to issues such as poor lift-to-drag ratios, low maximum lift coefficient values, and associated control issues. In order to produce enough lift for efficient flight, smaller flight vehicles may use entirely laminar sections or, like insects, may need to adopt innovative unstable aerodynamic systems. Miniature aircraft must operate in conditions more akin to those of little birds and giant insects than larger aircraft. Compared to a regular airplane in flight, the forces brought on by the air moving around the smaller objects are more evident. As a result, there is more drag, less lift at low speeds under the narrower wings, and less propeller efficiency. These aircraft are particularly vulnerable to rain, updrafts, and wind gusts. According to Robert Michelson of the Georgia Tech Research Institute (GTRI), miniature aircrafts employ novel, unorthodox technology, and scaling down conventional aircraft is not as straightforward as it might seem [1]. The devices are not just scaled-down versions of current pilotless military drones. But according to Dr. William R. Davis, manager of the MAV program at Massachusetts Institute of Technology's Lincoln Laboratory, nothing about creating microscopic air vehicles will be simple [2].

In addition to this, these sensors provide highly noisy measurements. Thus, the aerodynamic angles are to be estimated in a highly noisy environment, which poses another major challenge. One can identify measurement noise, process noise, bias or random walk and drift, etc., as the possible inconsistencies present in sensor data. Usually, the sensor measurements do not exactly produce the actual value but are dispersed around a mean value. However, the sensor noise characteristics are considered to have a zero mean and random in nature. Figures 11.2(a) and 11.2(b) shows the general normal distribution curve. A small value of σ indicates a high probability of the sensor measurement close to the actual value. Though it is a fact that a true Gaussian distribution can never occur in the real world, it is assumed that the sensor measurements follow such a distribution.

Further an important factor that imposes a greater challenge is that the signal conditioning. The measured sensor data has to be conditioned into a usable format. It could be any form of conditioning, from rectification to amplification to filtering. The information gathered from the sensors may lose certain important features. This has to be properly taken care and due attention is required in tackling this issue. One good example is the lag introduced due to the filtering process of the sensor information. Thus, one has to understand the significance of working with noisy sensor measurements and the proper signal conditioning aspects as highly important as the aerodynamic challenges involved (Figure 11.3).

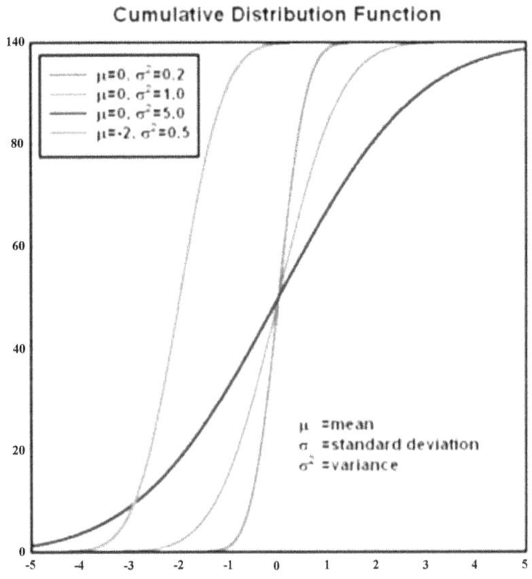

FIGURE 11.2(a) Cumulative distribution function.

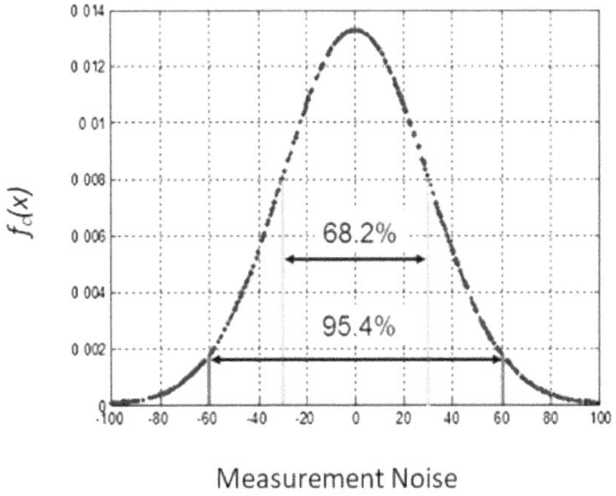

FIGURE 11.2(b) Gauss distribution curve.

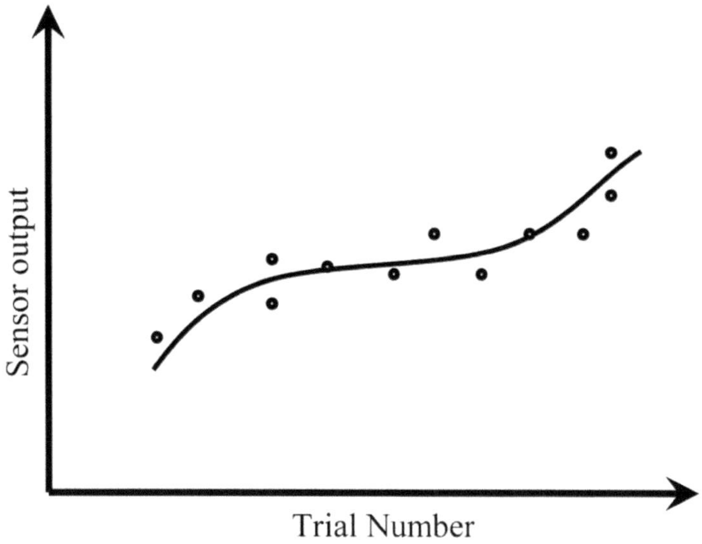

FIGURE 11.3 Noisy sensor measurements.

Missions involving healthcare and medical emergencies and related aspects demand good controller, which indeed require accurate feedback quantities. A notable work from William E. Green and Paul Y. Oh presents about the autonomous hovering of a fixed wing micro-air vehicle [1, 2]. They have developed *Proportional-Derivative* (PD) controller for the same. Another attractive research capability of UAVs is the cooperative missions. M.L. Cummings and P.J. Mitchell bring out the operational scheduling strategies for multiple UAVs [3]. Authors Marco Buschmann, Jens Bange and Peter Vorsmann talk about the application of MAV for meteorological purposes. They have implemented a feedback control structure with autopilot in the CAROLO aircraft [4].

11.4 NAVIGATION OF DRONES – FIXED WING AND MULTIROTOR

Navigation and guidance of drones is very vital aspect of their operation. Basics of flight mechanics throws light on how a flying machine is operated using transformations of the position and velocity about different frames of references. Nevertheless, this chapter is not going to emphasis on elaborating on these fundamentals' concepts. Yet we would like to provide a few necessary insights on various frames of references for the benefit of the readers.

11.4.1 FIXED WING AIRPLANES

As far as a fixed-wing drone system is concerned, they are operated mainly with the help of their wing aerodynamics. A drone systems flight is understood from the basic position vector notation and calculating their navigation using a few reference frames. A number of frames of references must be used for specifying relative positions and velocities, components of vectors and elements of matrices. In atmospheric flight, we commonly use Earth-fixed axes, vehicle-fixed axes, trajectory-fixed axes and atmosphere-fixed axes. It may be noted that various reference frames are reproduced from reference [1–3], and some of them are used in the present flight simulation and problem formulation.

11.4.1.1 INERTIAL REFERENCE FRAME – F_I

The inertial reference frame is a fixed frame of reference. Newton's second law is valid for the motion of a particle, in the sense that if 'f' is the sum of all external forces acting on the particle, and 'a' is acceleration relative to FI, then f = ma. The reference frame is taken to be either stationary or moving at a constant speed. Long-haul flights and interplanetary expeditions are the main uses of this frame of reference.

11.4.1.2 EARTH-FIXED REFERENCE FRAME – F_E

Any reference frame fixed to the Earth can be utilized as an inertial frame in many airplane dynamics issues, allowing for the ignoring of the Earth's rotation with respect to the inertial frame. As seen in Figure 11.4, two Earth-Fixed frames are of interest. A reference point on the equator and the Earth's axis are used to set the axis directions in the Earth-center frame, which is the first one. And the second one is the Earth-surface frame, with origin near the vehicle if possible and the z-axis directed vertically downwards. This frame of reference is applicable for short time missions.

11.4.1.3 VEHICLE-CARRIED VERTICAL FRAME – F_V

This is a reference frame in which the origin O_V is attached to the vehicle, usually at the mass center C, and the z-axis is directed vertically downwards along the gravity vector. In many applications the origin of the Earth-Fixed frame is close enough to the vehicle that the curvature of the Earth is negligible and F_V has axes parallel to F_E. Figure 11.5 shows the local and the vehicle carried reference frames. This reference frame captures only the translational motion.

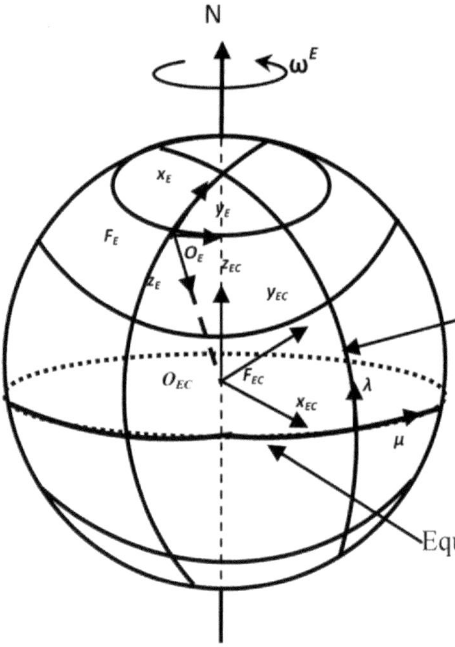

FIGURE 11.4 Earth axes.

11.4.1.4 ATMOSPHERE-FIXED REFERENCE FRAME – F_A

The relevant velocity of the atmospheric forces in atmospheric flight is that of the vehicle relative to the local atmosphere. So, it is essential to be concerned with the motion of the atmosphere. When the atmosphere is assumed to be at rest relative to the Earth, then F_A and F_E are the same. If the atmosphere is in uniform motion with respect to the F_E with velocity 'V_w,' then F_A is convicted relative to F_E with that velocity. The motion of F_A relative to F_E in this case is also a constant velocity 'V_w.' If the velocity of the vehicle mass center relative to F_A is denoted by 'V_a' so that its velocity 'V_e' relative to F_E is given as, sum of V_a and V_w.

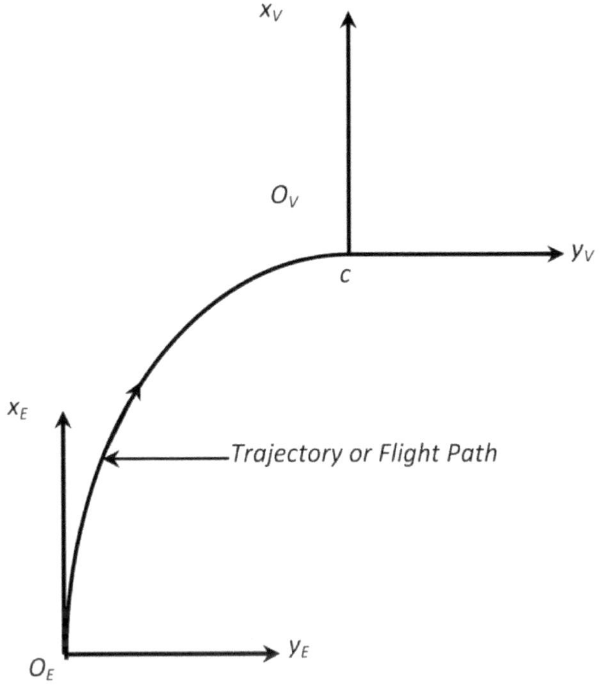

FIGURE 11.5 Local and vehicle carried vertical reference frames.

11.4.1.5 AIR-TRAJECTORY REFERENCE FRAME – F_W (WIND AXES)

This reference frame has its origin fixed to the vehicle at the center of mass and the x axis directed towards the velocity vector V of the vehicle relative to the atmosphere. If the atmosphere is at rest, then the origin would trace out the trajectory of the vehicle relative to the Earth and the x axis would always be tangential to it.

11.4.1.6 BODY-FIXED REFERENCE FRAME – F_B

Any set of axes fixed in a rigid body is a body-fixed reference frame. The origin of body axes is usually the mass center 'C.' Flight vehicles invariably have a plane of symmetry, and this plane is chosen to be C_{xz}, with z directed downward. The body axes are very important in-flight dynamics as shown in Figure 11.6. The angular velocity of F_B relative to F_I is ω (p, q, and r) and the components of V_B are u, v, *and* w.

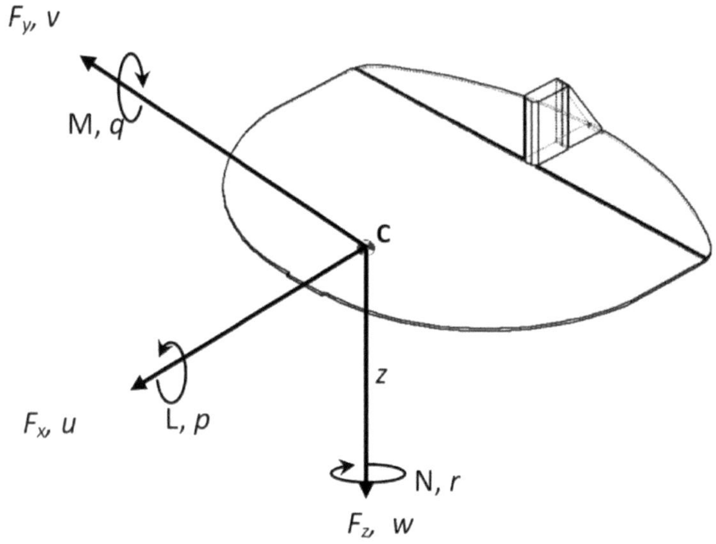

FIGURE 11.6 Vehicle body axes.

11.4.1.7 STABILITY AXES – F_S

This unique combination of axes is mostly used to examine minor deviations from a stable reference flight condition. F_S corresponds with the wind axis F_W under the reference condition if V is in the plane of symmetry, but it deviates from it during the disturbance and moves with the body. The x axis is chosen to lie on the projection of V in the plane of symmetry with the z axis also in the plane of symmetry if the reference flight condition is not symmetric, that is, with sideslip.

Figures 11.7 and 11.8 capture the aerodynamic quantities that are having pronouncing effects on the operational aspects of a fixed wing drone. Whereas the fixed wing vehicles exploit the aerodynamics, multirotors produce lift through their thrust forces produced by the motor-propeller systems. Such vehicles exhibit stability and VTOL capabilities and do not require long runways for their operations.

11.4.2 MULTIROTORS DRONES

Multirotor drones are extensively used in modern-day civil and healthcare missions. It is mainly due to the fact that these are stable vehicles and have the VTOL capability. These two features are definitely a great advantage for civilian applications. Multirotors are designed to suit the payload requirements in many different forms and configurations. Table 11.1 shows various configurations of multirotors based on the number of rotors. It ultimately boils down to the payload requirements for a particular application.

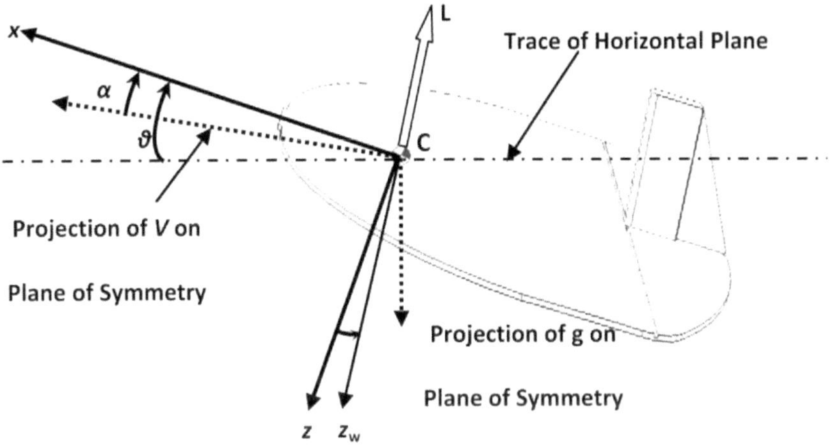

FIGURE 11.7 Plane of symmetry showing AOA.

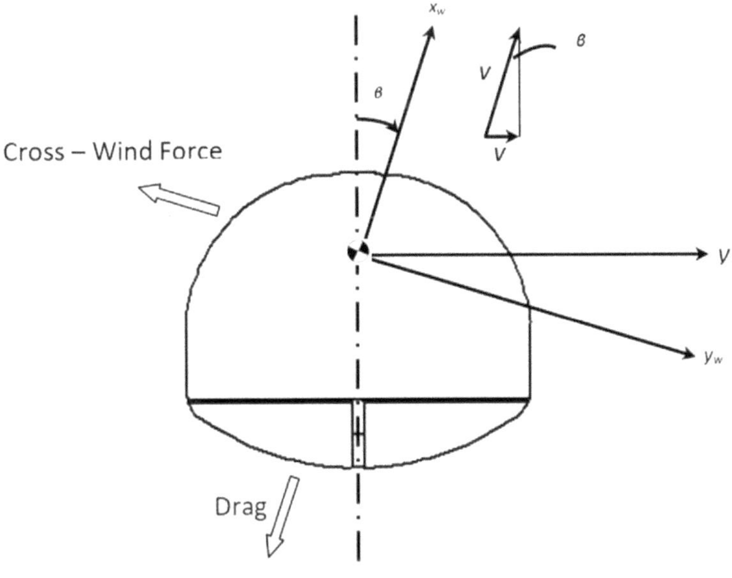

FIGURE 11.8 Drag and crosswind force vectors showing SSA.

TABLE 11.1 Multirotor Configurations

SL. No.	Multirotor Type	Payload
1.		Less payload
2.		Less payload
3.		Moderate payload
4.		High payload
5.	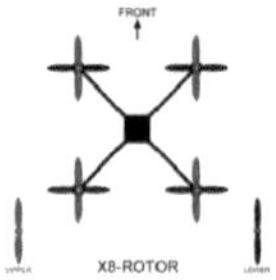	High payload
6.	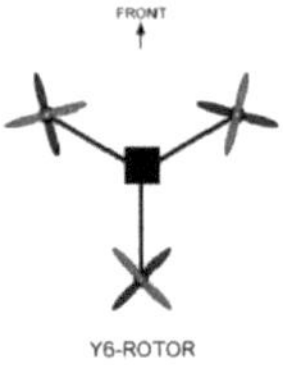	High payload

Thus, the term "multirotor" is a generic one which infers that it is a drone with multiple propellers for generating lift and also for the payload capacity. Figure 11.9 projects the axis system related to the body of a quadcopter configuration. The roll, pith, and yaw motions are shown with the notation's phi, theta and psi, respectively, similar to a fixed wing configuration.

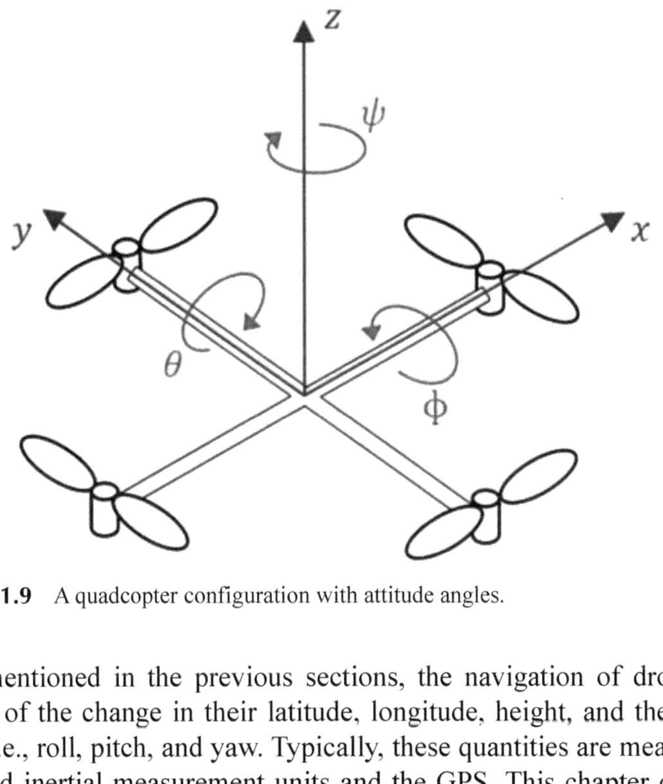

FIGURE 11.9 A quadcopter configuration with attitude angles.

As mentioned in the previous sections, the navigation of drones is a function of the change in their latitude, longitude, height, and the attitude angles, i.e., roll, pitch, and yaw. Typically, these quantities are measured by integrated inertial measurement units and the GPS. This chapter discusses the connection between these sensor measurements, navigation of drones applicable to medical and healthcare sector in the concluding remarks. Until now, we have seen the basic science behind drones and their operational requirements. In the next few sections, let us discuss the healthcare and medical applications of drone systems. It could be noticed that the main application of drone systems in healthcare is in the transportation domain. This chapter tries to provide a Birdseye view of such applications rather than attempting to focus on a particular scientific function for drone systems in healthcare.

11.5 MEDICAL APPLICATIONS OF DRONES

The idea of utilizing Drones in medical field for healthcare applications has evolved through a long way during the past decade. Drones are employed in a variety of end user applications as depicted in the block diagram shown in Figure 11.10.

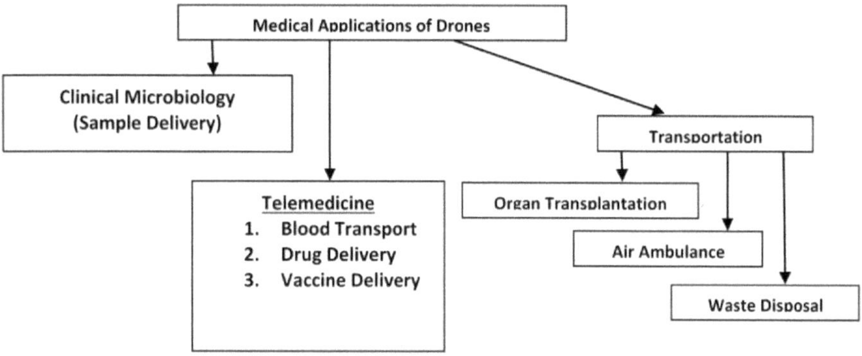

FIGURE 11.10 Block diagram showing medical applications of drones.

11.5.1 CLINICAL MICROBIOLOGY

Drones deployed for clinical microbiology analysis is a significant area considered in modern healthcare community. Collection of blood samples and transporting the same for testing and analyzing can be made effective and faster. In critical cases where time and efforts are vital factors in achieving tasks in remote and inaccessible areas, drone systems prove very effective [5–10]. It is helpful in identifying early infant infections of diseases such as HIV. In recent times, COVID testing has also been done very effectively using drones. It increases the reachability and access to the poor and under privileged people around the globe with a quick testing and treatment for highly infectious diseases. Such drones may be designed with lesser and medium payload capacities. The endurance and navigational issues must be tackled in order to employ them in remote areas as it may encounter loss of signal and GPS lag issues.

11.5.2 TELEMEDICINE AND BLOOD TRANSPORT

Drug delivery and blood transportations is another avenue for drones where they are seen as modern-day equipment with high efficiency. Researchers, manufacturers, and nonprofit groups are beginning to look to UAS to develop applications that raise efficiency and improve medical results since the potential are there. Recently, the first drone delivery from ship to shore in the United States was finished by drone company Flirtey. According to a news statement, the trip, which was organized in cooperation with the John Hopkins University School of Medicine and the nonprofit Field Innovation Team (FIT), showed how UAS can transport relief items like water and medical supplies in the event of a disaster. In the demonstration, drones were used to transport medical samples for emergency testing between an on-land medical relief camp in Cape May, New Jersey, and a test laboratory on a ship off the coast of New Jersey. Additionally, the drones moved medical supplies from the ship to the medical camp ashore [6–10].

Life-saving drugs and vaccines could be delivered more quickly and with greater accessibility thanks to drones. For instance, three years ago on the Pacific Island nation of Vanuatu, Wingcopter transported children's vaccines. In addition, we have tested potential regular supplies of insulin off the coast of Ireland. Currently, COVID-19 vaccines are sent through drones in Africa and Southeast Asia. Drones are a desirable platform for medical delivery because of their quick response times and capacity to traverse terrain that would otherwise be impenetrable. In 2007, scientists from Denel Dynamics and the National Health Laboratory Service (NHLS). The (UAV division) tested a proof-of-concept unmanned system to more efficiently transfer microbiological samples to NHLS facilities for fast Human Immunodeficiency Virus (HIV) testing from far-off clinics. The findings showed that drones can speed up diagnostics and facilitate medical decision-making. Drones can also be used to distribute important medications, such as anti-venom and other post-exposure vaccinations for canine and reptile bites are being developed in an effort to lower the incidence of casualties, which are especially common in rural regions.

The legal approval from aviation authorities is a significant barrier to the usage of medical drones. In India, using drones for commercial purposes is not allowed. According to Federal Aviation Administration (FAA) regulations, a license to fly a UAV is given in the US, however the drone must weigh less than 25 kg, stay within visual line-of-sight (VLOS), travel at a top ground speed of 100 mph, and soar to a maximum altitude of 400 feet.

11.5.3 TRANSPORTATION

Transportation is an exciting domain of application of drone systems. It includes transportation of medical equipment, organs for emergency transplantation, air ambulance and hazardous healthcare waste disposal. Drones may certainly be used to transfer medical devices and equipment to remote locations. It should be obvious that delivering some medical gadgets in an emergency in remote locations might be accomplished with ease using drones [11–18].

Drone monitoring of patients is reported recently to avoid direct intervention of the doctor to inspect a patient whose upper limb was transplanted. In addition to reducing the likelihood of a high rate of infection in patients undergoing organ transplantation, it has led to excellent surveillance. Organ transplantation surgeries are often assisted by drones in recent times. It is primarily because of the emergency situations and the troublesome road traffic conditions. Drone systems employed for organ transplantation deliver their services efficiently. But there are enormous challenges involved in deployment and operation of them. It includes design of the drone system itself due to the need for carrying organs at low temperatures. A refrigeration unit required for the organ transport is seen as the biggest contest in the operation of drones. This finally results in an increased payload requirement, increase in the number of motors and propellers to produce the required thrust and lift, increased battery capacity and also mechanisms to hold the refrigerated container for the organ. Such an application development is a potential area of research for the scientist in UAV community.

Another typical application is the air ambulance services. This application are must flourish in the long run because, ambulance services save lives trading off with time. Patients encountering need for emergency medical attention, especially accident victims, can be air-lifted as a general norm in the near future. Traffic conditions on road, in countries like India is densely hindering and highly congested resulting in loss of precious minutes of time drops. A drone ambulance is one of the best solutions to tackle emergency shifting of patients to hospitals. Yet, it has many operational inconveniences regulations, and also design constraints of the flying vehicle itself. Drone ambulance may not be completely unmanned; it needs basic medical professionals to be onboard. This is a huge constraint apart from carrying medical equipment and first aid devices [17–23].

The next imperious need for drones in healthcare sector is the handling of hazardous medical wastes. Waste management using drones has gained momentum in almost every field; but in public healthcare it is in a budding stage. Nevertheless, drones have huge potential for their deployment to segregate and eliminated medical wastes at dumpyards, chemical plants and other plants.

Drones can be deployed in surveillance of commercial areas and dump sites. Apart from medical wastes, these drones can detect general wastes, litterbugs and collect littered garbage in public areas such as parks and beaches. These drones can also keep an eye out for the illicit dumping of industrial trash. The recorded photographs and videos might be utilized to prove that illegal dumping practices are taking place. This is a tool that governments can employ to punish litterbugs and discourage illegal dumping and littering. Using drones to manage waste can greatly enhance several waste disposal processes. Drones have the ability to determine the compaction rate when surveying a landfill. The density of the waste inside each cell can be calculated using this information along with weighbridge data. This can be used by organizations to determine and compare the amount of waste present in each cell. By merging drone technology and IoT, these gadgets can improve their waste management capabilities even further. These techniques will enable drones to distinguish between recyclable and non-recyclable rubbish. Governments will be able to automate a number of waste management chores as a result. This will be a next-generation drone implementation that will help the world realize the operational wonders of drones.

11.6 CONCLUSION

Drones are flying machines with great deployment potential and uses for raising the standard of living for people. Their practical limitations, such as the sensor properties and GPS measurement lag problems, are discussed in the first portions of this chapter. As difficult as it is to operate a drone for any other scientific purpose, designing a drone for a particular healthcare use is a difficult task.

KEYWORDS

- **angle of attack**
- **intelligence**
- **proportional-derivative**
- **reconnaissance**
- **surveillance**
- **unmanned aerial vehicles**

REFERENCES

1. Dan, G., (2019). *The Drone Data Book* (pp. 1–253).
2. Wang, J. C., & Victor, C. Y. S., (2005). *Unmanned Aerial Vehicle, Development Trends and Technology Forecast, DST Horizons* (pp. 1–18).
3. Thomas, J. M., (2009). On the birth of micro-air vehicles. *International Journal of Micro-Air Vehicles, 1*(1), 1–11.
4. Maziar, A., (2007). *Classification of Unmanned Aerial Vehicles*. Lecture Notes, University of Adelaide.
5. Thomas, J. M., (1999). *Aerodynamic Measurements at Low Reynolds Numbers for Fixed Wing Micro-Air Vehicles*. Special Course on Development and Operation of UAVs for Military and Civil Applications, VKI, Belgium.
6. Rosser, Jr. J. C., Vignesh, V., Terwilliger, B. A., & Parker, B. C., (2018). Surgical and medical applications of drones: A comprehensive review. *JSLS, 22*, e2018.00018.
7. Dekoulis, G., (2018). *Introductory Chapter: Drones, Drones E Applications*. IntechOpen.
8. US Department of Defense, (2005). *Unmanned Aerial Vehicles Roadmap*, 2005e2203 Washington, DC: Department of Defense, Office of the Secretary of Defense.
9. Greaves, R. F., Bernardini, S., Ferrari, M., Fortina, P., Gouget, B., Gruson, D., et al., (2019). Key questions about the future of laboratory medicine in the next decade of the 21st century: A report from the IFCC-emerging technologies division. *Clin. Chim. Acta, 495*, 570e89.
10. Ackerman, E., & Koziol, M., (2019). *In the Air with Zipline's Medical Delivery Drones*. IEEE Spectrum. Available at: https://spectrum.ieee.org/robotics/drones/in-the-air-with-ziplines-medical-delivery-drones (accessed on 20 November 2023).
11. Stoakes, U., (2015). *The Leapfrog Opportunity in the World's Underserved Healthcare Markets.* Forbes.
12. United Nations Department of Economic and Social Affairs, (2018). *World Economic and Social Survey*. Available online: https://www.un.org/development/desa/dpad/wp-content/uploads/sites/45/publication/WESS2018_full_web.pdf (accessed on 20 November 2023).
13. Glauser, W., (2018). *Blood-Delivering Drones Saving Lives in Africa and Maybe Soon in Canada.* Available online: http://www.cmaj.ca/lookup/doi/10.1503/cmaj.109-5541 (accessed on 20 November 2023).
14. Carrillo-Larco, R. M., Moscoso-Porras, M., Taype-Rondan, A., Ruiz-Alejos, A., & Bernabe-Ortiz, A., (2018). The use of unmanned aerial vehicles for health purposes: A systematic review of experimental studies. *Glob. Health Epidemiol. Genom., 3*, e13, 1–10. doi: 10.1017/gheg.2018.11. Available online: https://www.cambridge.org/core/product/identifier/ (accessed on 20 November 2023).
15. Hampson, M., (2019). Drone delivers human kidney: The organ was flown several kilometers by a drone without incurring damage. *IEEE Spectr., 56*, 7–9. Available online: https://ieeexplore.ieee.org/document/8594776/ (accessed on 20 November 2023).
16. Lippi, G., & Mattiuzzi, C. *Biological Samples Transportation by Drones: Ready for Prime Time*. http://www.ncbi.nlm.nih.gov/pubmed/27047951 (accessed on 20 November 2023).
17. Al-Zayer, M., Trelligus, S., Bhandari, J., Dave, F. S., & Folmer, E., (2016). Exploring the use of a drone to guide blind runners. In: *ASSETS 2016: Proceedings of the 18th International ACM SIGACCESS Conference on Computers ACM* (pp. 263–264). New York, NY, USA.

18. Scott, J. E., & Scott, C. H., (2017). Drone delivery models for healthcare. *Proceedings of the 50th Hawaii International Conference on System Sciences* (pp. 3297–3304). Village, HI, USA.
19. Thiels, C. A., Aho, J. M., Zietlow, S. P., & Jenkins, D. H., (2015). Use of unmanned aerial vehicles for medical product transport. *Air Med. J.*, *34*, 104–108.
20. Canadian Agency for Drugs and Technology in Health, (2018). *Health Technology Update: A Newsletter on New and Emerging Healthcare Technologies in Canada Rural and Remote Issue*.
21. Mendez, I., Jong, M., Keays-White, D., & Turner, G., (2013). The use of remote presence for healthcare delivery in a Northern Inuit Community: A feasibility study. *Int. J. Circ. Health*, *72*, 21112.
22. Arksey, H., & O'Malley, L., (2005). Scoping studies: Towards a methodological framework. *Int. J. Soc. Res. Methodol Theory Pract.*, *8*, 19–32.
23. Hsieh, H. F., & Shannon, S. E., (2005). Three approaches to qualitative content analysis. *Qual. Health Res.*, *15*, 1277–1288.

CHAPTER 12

ADVANCED MECHANICAL VENTILATOR

U. NITHISHKUMAR, S. SENDIL KUMAR, J. MEGHALA, and J. SUNDARAPERUMAL

Department of Electrical and Electronics, S. A. Engineering College–Autonomous, Chennai, Tamil Nadu, India

ABSTRACT

This study demonstrates the design of a sophisticated mechanical ventilator. The unavailability of mechanical ventilators for COVID-19 patients motivated us to build this sort of ventilator—the COVID-19 pandemic has been very devastating in certain areas, notably among the poor. A mechanical ventilator is a piece of equipment that helps a patient puff (ventilate) when they are undergoing surgery or experiencing breathing difficulty due to a serious disease. An empty channel (fabricated airway) is put into their mouth to connect the patient to the ventilator, and also below into their prime airway or trachea. When patients are unable to breathe on their own, they will be placed on a ventilator. Mechanical ventilation is the most often used supportive technique in critical care hospitals. External respiratory support, as well as access to lower airways through tracheostomy or endotracheal tubes, has long been used to aid the failing respiratory apparatus. The purpose of building a better mechanical ventilator is to mitigate the effects of global scarcity and to limit the spread of COVID-19. AMBU bag automation is used in this procedure. This design not only benefits the patients, but it also provides a healthy environment for them. This design will assist to reduce the demand for extra personnel as well as the spread of new illnesses. The LCD monitor displays the volume of air delivered to the patient, the number of breaths per minute, SPO_2, BPM, O_2%, and CO_2%. This revolutionary mechanical ventilator is utilized to clean and sanitize the patient's

Robotics and Automation in Healthcare: Advanced Applications. R. Thanigaivelan, Sanjay Singh, & Clement Christy Deepak C. (Eds.)

surroundings. This ventilator, which we designed and built with Arduino, fits all of the criteria for developing a long-lasting but modern ventilator to help in pandemic scenarios.

12.1 INTRODUCTION

COVID-19 is a dreadful sickness brought on by a virus that has altered our life. Many individuals perished as a result of this ailment due to a shortage of medical care. It also affects our lungs. Mechanical ventilation is required if a patient suffers respiratory failure. All standard techniques may be used to differing degrees for this. Although this appears to be reasonable, there is no clear consensus on when, how, and how much of the patient's breathing work should be decreased. Inadequate support may result in stomach discomfort or weakness, requiring the activation of auxiliary inspiratory muscles, which can lead to respiratory acidosis. Excessive assistance might result in respiratory alkalosis and decreased respiratory drive, which can result in patient-ventilator asynchronies and sleep disruptions. In 1929, mechanical ventilation was introduced. The first documented usage of ventilators dates back to biblical times. The first reported use of negative-pressure ventilation occurred in the 1800s. Definite force devices began towards to see spurge as pneumatic compensating devices in medical center, health center, and intensive care units (ICU) in the 1900s [1], but this technology did not begin to be developed until the 1940s [3]. Since then, there has been a significant increase in the production of positive mechanical ventilators, which are routinely used in critical care units. Each industrial generation contains characteristics that distinguish it from previous and later generations. These generations' great industrial growth laid the basis for today's smart generation. In the 1950s, the first generation of critical care facility ventilators appeared. It just regulated the volume of air [2, 4]. The ventilation device was a basic electrical circuit with a compressor, and the respiratory rate and tidal volumes were automatically determined.

The first-generation ventilators, Additional alarms, such with elevated pressure, high respiratory rate, or short tidal size, were introduced to the second generation of critical care ventilators.

The notion of needed intermittent ventilation was also pioneered in adult ventilation, as was the method of regulating air and oxygen pressure using a Servo came in medical practice [6]. In the late 1970s, the notion of feedback-control ventilation first arises [3, 7].

The third generation of critical care ventilators, such as the Servo 300 ventilators, used the microprocessor idea to regulate the mechanical ventilation that came in the third generation. This generation of devices enabled monitoring of the patient's state as well as each restriction volume of the ventilator instrument [13].

The fourth generation of critical care units, the most recent instrument was employed [14]. Mechanical ventilator is classified into two types: positive-pressure ventilation and pessimistic-force ventilation. In the negative pressure ventilation, it draws air into the lungs through contracting and expanding the chest.

Initially the ventilators we used negative-pressure type ventilator they are tiny in usage.

1. **Iron Lung:** The first mechanical ventilator, a piston chamber completely encased the patient till the neck.
2. **Chest Cuirass:** A negative pressure is created by using the thin shell that is attached to the patients chest.

12.1.1 POSITIVE-PRESSURE VENTILATION

This type of ventilator was created in the early 1950s to treat polio victims with respiratory failure. These ventilators provide air to the patient's lungs via a tube. It has both invasive and non-invasive features.

12.1.2 INVASIVE VENTILATION

Invasive ventilation comes in two varieties. Endotracheal intubation and Tracheostomy are two of them. Endotracheal intubation involves simply connecting an air hose to the patient's lungs through the mouth. Tracheostomy is nothing more than a procedure in which surgeons perform surgery on a patient's neck area and attach an air tube to it.

12.1.3 NONINVASIVE VENTILATION

We can make advantage of at home, noninvasive mechanical ventilators come with masks. Noninvasive mechanical ventilators are classified into three types:

1. **Continuous Positive Airway Pressure (CPAP):** It maintains continuous and consistent air pressure in the lungs.
2. **Autotitrating (Variable) Positive Airway Pressure (APAP):** It regulates air pressure to allow for normal breathing.
3. **Bilevel Positive Airway Pressure (BiPAP):** It carries air at two pressure levels for inhalation and exhalation.

This design is based on manual BVM automation (AMBU-bag). When the bag is squeezed, air enters the patient's lungs, and the nonreversible breathing valve prevents backfiring. The AMBU bag then spreads itself by sucking air through the valve on its backside. An oxygen cylinder or ambient air can be used as "fuel." In the latter case, a tank can be attached to collect any excess oxygen that the patient did not use. The vast majority of DIY ventilators on the market use AMBU bags and a pumping mechanism. This streamline pumping operations while also increasing the user interface of the machine. The fundamental component of this design is an Anaglyph printed rod connected to a servo motor that compresses the AMBU bag. There is also a control panel for precisely adjusting the ventilation settings. Variables can be used to alter the ventilation settings the tidal volume and rate of respiration. The robotic arm in this mechanical ventilator is used to protect the oxygen mask near the patient's face. The robotic arm is attached to the mechanical ventilator's base. Mobile devices are used to operate the robotic arm. The DC motor is connected to a lead screw. The traveling nut secures the oxygen mask to the patient's face while the motor turns. In this configuration, an inverter is installed for emergency power backup. The L298N driver is used to control the four DC motors. The microcontroller is linked to the driver. The mobile devices connected to the microcontroller through Bluetooth module allow the automated base to be moved anywhere. The wheels are connected to the DC motor at the ventilator's base. When the mechanical ventilator is not in use, this device may be used to clean the flooring.

12.2 EXISTING WORKS OF THE VENTILATORS

A mechanical ventilator is an apparatus that supports patients in breathing (ventilating) though they are undertaking surgery or are incapable to breathe on their individual due to a severe disease. A hollow tube (artificial airway) is introduced into the patient's mouth and below into their basic airway or trachea where the ventilator is link to them. They keep on the ventilator till they build on sufficient to breathe by their own. A mechanical ventilator is used to cut the labor of breathing till the patient improves adequately to no extensive they need it. The equipment ensures that the body receives enough oxygen. This is required when some disorders make regular breathing impossible.

COVID-19 is mostly spread through close contact with patients. The infection spreads to doctors and other patients in hospitals. To address these shortcomings, we developed a project on sophisticated mechanical ventilators. This idea might be implemented in ICUs and COVID patient units using mobile devices. Our model has the ability to clean and sanitize the floors. As a result, disease transmission and personnel may be limited. The robotic arm is used to affix the patient's oxygen mask on his or her face. We've added a converter design that will come in helpful during power outages.

12.3 PROBLEM STATEMENT

- After our survey, there is a need for mechanical ventilators in India to combat COVID 19.
- To manage patients at that period. One of the most serious difficulties is that the sickness has spread to the clinicians who are treating COVID-affected patients.
- As a result, we chose the title "advanced mechanical ventilator" for our research.

12.4 RECENT ADVANCES IN THE FIELD

It has been designed and developed to provide helpful options and improved interfaces. In recent years, the utility of processes and symptoms for HFNC oxygen and conventional NIV in acute hypoxemic nonhypercapnic respiratory failure has been enhanced. In one randomized controlled trial, intubation amounts were comparable in patients receiving HFNC, NIV managed via facemask, and standard oxygen therapy; however, mortality rates were lower with HFNC than with NIV or standard oxygen therapy, and HFNC was associated with improved comfort and less dyspnea [9]. Patients affected with ARDS who were randomly assigned to obtain NIV over helmets or face masks needed lesser respiratory rates, smaller intensive care unit halts, and decreased humanity [10].

There has been a growing awareness and understanding of the hazards involved with mechanical ventilation in recent years. Ventilatory techniques that use short end-inspiratory plateau airway pressure, short tidal volume (VT), and increased PEEP improve ARDS survival. Based on the limited epidemiologic data collected in the setting of controlled tidal capacity and no impulsive breathing, dynamic pressure has remained recognized as a major cause of death. The observational LUNG SAFE study, which looked at how clinicians ventilate ARDS patients, discovered that less than two-thirds of patients had V T of less than 8 ml/kg of frame weight and 82.6% had PEEP less than 12 cm H_2O. Increased survival was fully related to advanced PEEP, lower peak, plateau, and heavy pressures, and a lower respiratory rate [13]. Specialized ventilators, such as neonatal ventilation, grown ventilation, NIV, and carrying, will become obsolete. The next generation of ICU ventilators will be able to handle the full scenario as well as or better than earlier single ventilators. According to the research, certain ventilators are already capable of providing ventilation in a range of situations, and that all will be in the future [12]. Protocols will become an essential component of the ICU ventilator's basic operation. As more information on providing lung-protective ventilation and treating specific diseases becomes available, ventilators will be able to incorporate signal-based algorithms into their basic operating strategy. In addition to absolute volume, the forthcoming ventilator will need to enter the patient's elevation and gender, as well as values in mL/kg approximated to body load. The acute respiratory distress syndrome network protocol, as well as other ways for performing lung-recruitment exercises and setting PEEP, will be used on future ventilators. These techniques will still need the doctor to select fundamental settings, however the ventilator will have control to ensure that ventilation for a detailed sickness condition remains take out in line with the most recent evidence-based standards.

12.5 BLOCK DIAGRAM

Figure 12.1 shows the block diagram for impulsive ventilator. The list of hardware requirements to build mechanical ventilators is given below:

- Arduino Uno R3;
- Servo motor;

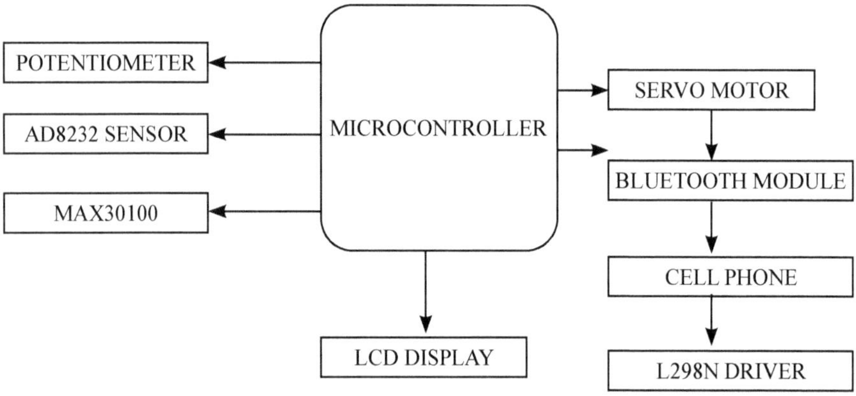

FIGURE 12.1 Block diagram.

- Potentiometer;
- L298n driver;
- MAX30100 pulse oximeter;
- AD8232 sensor;
- I2c LCD display.

Figure 12.2 shows the circuit diagram for the block diagram shown in Figure 12.1. Arduino is connected to potentiometer to 1 and 2 servo motor is connected to Arduino and LCD display is also linked to Arduino.

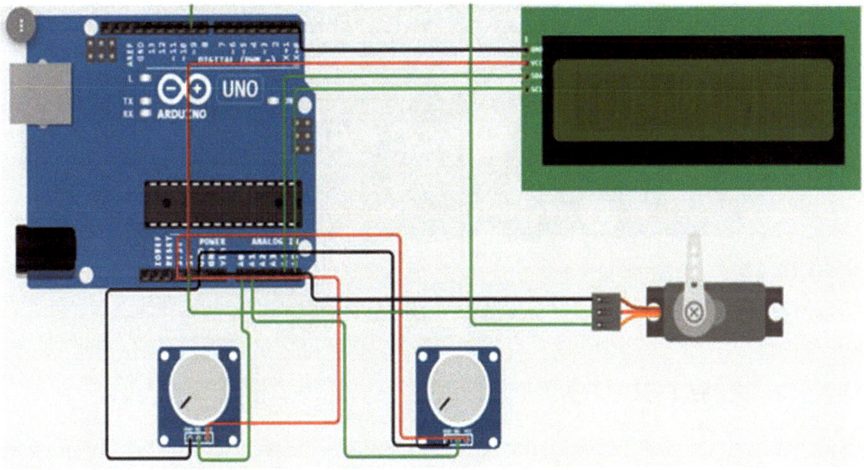

FIGURE 12.2 Circuit diagram.

12.5.1 ARDUINO UNO R3

Figure 12.3 shows the Arduino UNO R3 I which it consists of it is ATmega328-based Arduino Uno is a one chip computer board. It contains 14 digital I/O pins, six analog inputs, a 16 MHz crystal oscillator, a power connector, a USB connection, an ICSP header, and a reset button. It includes everything you need to get started with the one-chip computer; simply connect it to a computer by USB or power it with an AC-to-DC converter or battery. Previous versions of the Arduino software are evolving at the same rate as the current ones. It also includes the Uno board.

FIGURE 12.3 Arduino UNO.

Image credit: oomlout. https://creativecommons.org/licenses/by-sa/2.0/

12.5.2 RESET BUTTON

Similar to the new Nintendo, the Arduino features have reset button. By pressing it, the reset pin is momentarily connected to earth, resetting some code that is layered arranged in the Arduino. This might be quite beneficial if your code does not rehash yet you need to test it several times. In any case, unlike the original Nintendo, blowing on the Arduino frequently resolves any issues.

12.5.3 SERVO MOTOR

Figure 12.4 shows the servo motor used in the Arduino board. A servo motor is a rotating actuator that allows for correct angular position control. It is completely comprised of a motor and a sensor for station feedback. A servo drive is required to complete the system. Servo motors come in a range of sizes and three primary types. Positional rotation, continuous rotation, and linear rotation are the three types. Servos with positional rotation rotate 180°. They also include brakes on the gear system to protect the output shaft from spinning too rapidly.

FIGURE 12.4 Servo motor.

12.5.4 POTENTIOMETER

Figure 12.5 shows the potentiometer. The potentiometer is a device used to measure unknown voltages by differentiate them to familiar voltages.

Potentiometer are used to compute the EMF and internal resistance of a particular cell, also to match the EMF of multiple cells and used to quantity the displacement at any direction. Linear potentiometers measure movement linearly, whereas rotary potentiometers are used to measure the rotational displacement and regulate the amount of air and respiratory rate in a mechanical ventilator.

12.5.5 L298N DRIVER

Figure 12.6 shows the L298N Motor Driver Module is available in high power and is used to power both DC motors and stepper motors. This module employs a 78M05 5 V regulator and an L298 motor driver IC. The L298N Module is suitable for controlling up the four DC motors or two DC motors with viewpoint and permitted speed. Supply voltage of this driver module is 5 V to 35 V range and can convey 2 A of continuous current per channel.

FIGURE 12.5 Potentiometer.

Source: Public domain.

FIGURE 12.6 L298N motor driver.

12.5.6 MAX30100 PULSEOXIMETER

Figure 12.7 shows MAX30100 is a sensor system used for pulse oximetry and to monitor heart rate.

FIGURE 12.7 MAX30100 pulse oximeter.

The pulse oximetry and heart rate signals are detected by two LEDs. Signal processing is accomplished by the use of a photodetector, better optics, and low-noise analog. This sensor may be used in conjunction with any single-chip computer, such as Arduino, ESP8266, or ESP32, to measure the patient's health parameters. The operational voltage of the sensors is 1.8 V and 3.3 V sources, which may be turned down by software with a modest standby current, allowing the power supply to be connected at all times.

12.5.7 AD8232 SENSOR

Figure 12.8 shows the AD8232 sensor features a rapid restoration function that reduces restoration time. Otherwise, the settling tails of the high-pass filters are rather long.

After an abrupt signal shift that supports the amplifier, this sensor automatically adapts to a higher filter cutoff (such as a lead away situation). The IC of the Heart Rate Monitor contains nine connections to which you may add pins, cables, or other connectors. SDN, LO^+, LO^-, OUTPUT, 3.3 V, and GND are important pins to use with an Arduino or other programming board. It has a 3.3 V operating voltage.

12.5.8 I2C LCD DISPLAY

Figure 12.9 shows the I2C LCD is a simple display module that can help with display. It can lessen the difficulties of making, allowing creators to focus on the heart of the task.

FIGURE 12.8 AD8232 sensor.

FIGURE 12.9 I2C LCD display.

We created the Arduino library for I2C LCD, which allows users to accomplish complicated visuals and text output skills with limited lines of code. This I2C 16×2 Arduino LCD Screen employs an I2C communication interface, which requires four pins from microcontroller to function. This type of library will save at least four Arduino digital/analog pins. Entire connections are XH2 standard (Breadboard type). The I2C protocol is used to connect two or more integrated circuits (Integrated Circuits).

12.5.9 METHODOLOGY

- At initial we use MAX30100 pulse oximeter to measure the saturation of oxygen and BPM of the patient.

- By the result of pulse oximeter, the tidal volume and respiratory rate is set using the potentiometer.
- By getting the feedback from the micro-controller, the servo meter is used to pump the AMBU bag.
- AD8232 ECG sensor is used to measure the patients ECG and that is displayed using the monitor screen.

12.5.10 RESULT OBTAINED FROM THE VENTILATOR

Figure 12.10 shows the pumping method this concept is based on automating an AMBU bag using a servo motor connected to a microcontroller.

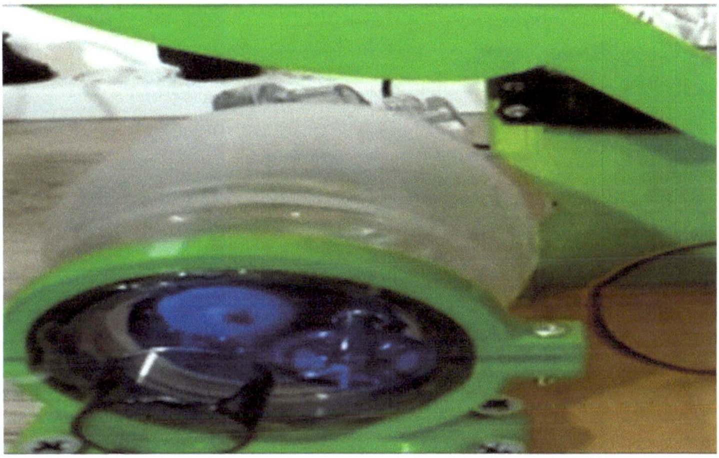

FIGURE 12.10 Pumping method.

The servo motor is linked to the AMBU bag through a rod-like mechanism. The air that comes out of the AMBU bag is sent to the patients via the respiratory circuit.

Figure 12.11 shows the simulation output from the LCD display. The LCD monitor displays shows the respiratory rate 20 and tidal volume as 60.

According to patient condition, the BPM and SpO_2 measure the value using the MAX3010 sensor and display using the LCD display both may be controlled using the potentiometer. On the monitor screen, the patients' ECGs may be observed. This mechanical ventilator's robotic arm is used to affix the oxygen mask to the patient's face.

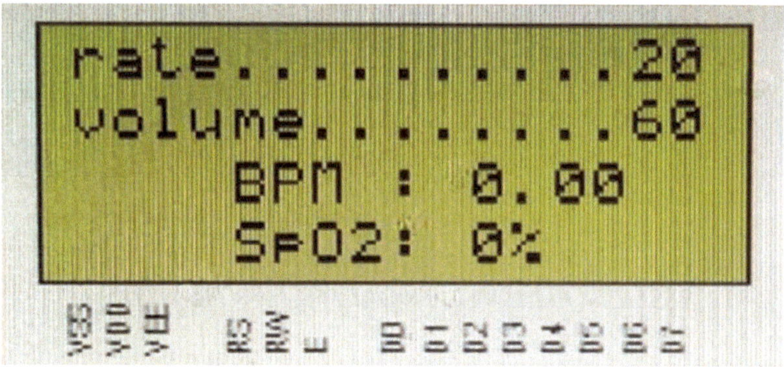

FIGURE 12.11 LCD display with simulation output.

Figure 12.12 shows the ECG simulation output of the Ventilator. The robotic arm is hooked to the base of the mechanical ventilator. The robotic arm is controlled via mobile devices. A lead screw is linked to the DC motor. While the motor is turning, the traveling nut binds the oxygen mask to the patient's face. An inverter is placed in this design for emergency power backup. The one-chip computer is utilized to automate the basis of the mechanical ventilator (Arduino). L298N driver is connected with the 4 DC motors. The driver is attached to the microcontroller. The automated base may be relocated anywhere thanks to mobile devices connected to the microcontroller through Bluetooth module. At the ventilator's base, the cleaning mechanism is linked to the dc motor. This gadget may be used to clean the flooring while the mechanical ventilator is not in operation.

12.6 ADAPTABILITY

- This approach is useful for transferring patients from ICU or home to ambulance, or at any accident scene.
- It assists those who administer first aid by using a methodical and simple approach; even inexperienced people may use it.
- Simple to reprogram if system modifications are necessary.

12.7 CONCLUSION

As a consequence, our mechanical ventilator is utilized to assess the patient's SpO_2 and BPM so that the respiratory rate and tidal volume may be adjusted.

When there are in need of the ventilator. Our ventilator also has a capable of measuring the ECG of the patient when they are in the ventilator. The new features that we installed in our ventilator compared to the current ventilator in the market are it can be used to clean and sanitize the area around the patients when it is not in used also the robotics arm is used to fix the oxygen mask to the patent face.

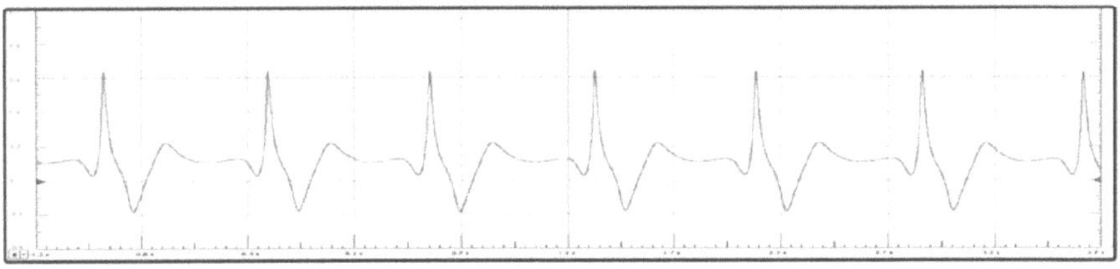

FIGURE 12.12 ECG simulation output.

ACKNOWLEDGMENT

We thank S.A. Engineering College for providing resources and encouraging us to do this project.

KEYWORDS

- **auto-titrating positive airway pressure**
- **bilevel positive airway pressure**
- **continuous positive airway pressure**
- **COVID-19**
- **intensive care units**
- **invasive ventilation**
- **short tidal volume**

REFERENCES

1. Claus, F. V., Gerard, J. C., Fernando, J. M., Antonio, A., Peter, J. B., Jean, B., Bartolome, R. C., et al., (2017). Global strategy for the diagnosis, management, and prevention of chronic obstructive lung disease 2017 report. GOLD executive summary. *Am. J. Respir. Crit. Care Med., 195*(5), 557–582. doi: 10.1164/rccm.201701-0218PP.

2. Rochwerg, B., Brochard, L., Elliott, M. W., Hess, D., Hill, N. S., Nava, S., Navalesi, P., et al., (2017). Official ERS/ATS clinical practice guidelines: Noninvasive ventilation for acute respiratory failure. *Eur. Respir. J., 50*(2).
3. Emad, M., Khaled Al, A., Husham, F., Elamir, A., Victor, V. M., Abdulwahab, A., & Hasan, A., (2022). Designing hybrid mechanical ventilator system based on Arduino and raspberry Pi 4. *Journal of Medical Devices, 16*(1), 014501–014508. doi.org/10.1115/1.4054036.
4. Simonds, A. K., & Elliott, M. W., (1995). Outcome of domiciliary nasal intermittent positive pressure ventilation in restrictive and obstructive disorders. *Thorax, 50*(6), 604–609. doi: 10.1136/thx.50.6.604.
5. Kacmarek, R. M., (2011). The mechanical ventilator: Past, present, and future. *Respir. Care, 56*(8), 1170–1180. doi.org/10.4187/respcare.01420.
6. Cook, T. M., & Kelly, F. E., (2015). Time to abandon the 'vintage' laryngeal mask airway and adopt second-generation supraglottic airway devices as first choice. *Br. J. Anaesth., 115*(4), 497–499. doi: 10.1093/bja/aev156.
7. Platen, P. V., Pomprapa, A., Lachmann, B., & Leonhardt, S., (2020). The dawn of physiological closed-loop ventilation: A REVIEW. *Crit. Care, 24*(1), 1–11. doi: 10.1186/s13054-020-2810-1.
8. Chopin, C., & Chambrin, M. C., (1993). Closed-loop control in mechanical ventilation. *Yearbook of Intensive Care and Emergency Medicine 1993* (pp. 499–507). Springer, Berlin, Heidelberg. doi: 10.1007/978-3-642-84904-6_46.
9. Tobin, M. J., (2001). Advances in mechanical ventilation. *New Engl. J. Med., 344*(26), 1986–1996. doi: 10.1056/NEJM200106283442606.
10. Frat, J. P., Thille, A. W., Mercat, A., Girault, C., Ragot, S., Perbet, S., Prat, G., et al., (2015). High-flow oxygen through nasal cannula in acute hypoxemic respiratory failure. *N Engl. J. Med., 372*, 2185–2196.
11. Patel, B. K., Wolfe, K. S., Pohlman, A. S., Hall, J. B., & Kress, J. P., (2016). Effect of noninvasive ventilation delivered by helmet vs face mask on the rate of endotracheal intubation in patients with acute respiratory distress syndrome: A randomized clinical trial. *JAMA, 31*(5), 2435–2441.
12. Laffey, J. G., Bellani, G., Pham, T., Fan, E., Madotto, F., Bajwa, E. K., Brochard, L., et al., (2016). Potentially modifiable factors contributing to outcome from acute respiratory distress syndrome: The lung safe study. *Intensive Care Med., 42*, 1865–1876. doi: 10.1056/NEJMoa1503326.
13. Ferreira, J. C., Chipman, D. W., Hill, N. S., & Kacmarek, R. M., (2009). Bilevel vs. ICU ventilators providing noninvasive ventilation: Effect of system leaks: A COPD lung model comparison. *Chest, 136*(2), 448–456.

CHAPTER 13

DEVELOPMENT OF A MEDICAL DECISION SUPPORT SYSTEM FOR EARLY DETECTION OF CARDIAC DISORDERS

S. SUMATHI[1] and S. PONLATHA[2]

[1]*Department of Electrical and Electronics Engineering, Mahendra Engineering College, Namakkal, Tamil Nadu, India*

[2]*Department of Electronics and Communication Engineering, Mahendra Engineering College, Namakkal, Tamil Nadu, India*

ABSTRACT

This chapter explains how to reduce ECG noise and quickly find QRS complex detection using cubic spline waveform technology. This clinical application is best suited for the suggested simulation since it lowers the likelihood of error. Cardiac problems are automatically recognized and monitored using the derived ECG features. Five different heartbeat types were studied using data from the MIT-BIH database. The DNN system has a 99.5% classification accuracy overall. In the future, the multi-WT and singularities detection approach would be used to retrieve the ECG characteristic points. The simulation findings are very trustworthy, and further assistance in identifying cardiac problems at every stage might be given by the classifier.

Robotics and Automation in Healthcare: Advanced Applications. R. Thanigaivelan, Sanjay Singh, & Clement Christy Deepak C. (Eds.)

13.1 INTRODUCTION

In this competitive and flourishing technological advancement occurring across the globe, there is an essential need for the evolution of advanced concepts in the area of biomedical engineering. Being the forerunner among the emerging areas in science and technology, this field demands extensive research. Worldwide, heart disease, stroke, and other forms of cardiovascular disease (CVD) generate over 17 million annual deaths. The World Health Organization (WHO) predicted that 16.7 million people around the world died unexpectedly from CVD in 2003. Heart disease and stroke will replace cancer as the world's leading causes of mortality and morbidity by 2020, with more than 20 million people expecting to experience these conditions each year by 2030 (Atlas of Heart Disease and Stroke, WHO 2004). The Eastern Mediterranean Region will see the biggest percentage increase. The South-East Asia region will see the most extensive spike in fatalities. The primary goal of this chapter is to implement MDSS system for Ultrasound Common Carotid Artery (CCA) and identify raw electrical pulses that are governing components of raw ECG play a vital role in diagnosis and further processing. Enormous electrical equipment and electrical expertise are essential in understanding processing, its execution, and various allied mechanisms. The heart and the corresponding ECG are found to draw a drastic focus from an electrical discipline point of view. Detecting cardiovascular illness early can be the difference between life and death. This chapter is made to partially extend a valuable contribution to the field of Biomedical Engineering.

13.2 ECG SIGNAL

The electrical signal flow in a human body surface generated by the heart is illustrated by an ECG. The ECG signals generated by elegant currents in the heart's tissues. The ECG waveform has a number of points of reference, including P, Q, R, S, T, and U in Figure 13.1.

The characteristics that determine to choose whether or not advise the ECG signs of heart disease can be determined by physicians. A single waveform begins at the baseline and develops into either waveform afterwards. Effective deflections are defined as the part of the ECG waveform above the baseline. Even though the U wave's origin is uncertain, it is believed that little U waves and large U waves have various physiological origins. U waves typically have an amplitude which is less than one third that of the T wave behind them. The use of certain signals is fundamental for the

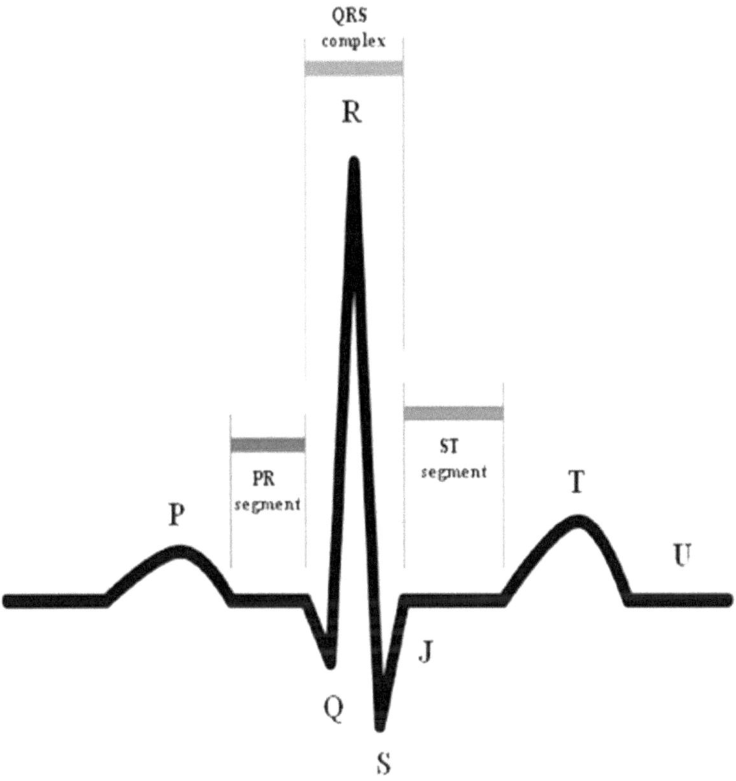

FIGURE 13.1 A typical normal ECG waveform.

analysis of heart problems. In recent years, computer-aided analysis was used to interpret and characterize ECG cardiac arrhythmias. The major responsibility in evaluating the ECG is R Peak determination. In recent years interpretation and classification of ECG cardiac arrhythmias using computer aided analysis. The R Peak detection is the major roll to analyze the ECG and easy to identify the cardiac abnormalities. In recent years, computer-aided analysis has been used to interpret and characterize ECG cardiac arrhythmias. The R Peak detection plays an important role in the analysis of the ECG and makes it simple to spot cardiac abnormalities. In clinical practice, the electrocardiogram (ECG), which illustrates the electrical activity of the heart, is commonly implemented. After a patient exercise during a physical examination in a hospital, the doctor records the ECG to determine the patient's cardiac health. The Holter ECG recording equipment is most extensively employed. When monitoring a patient's ECG to recognize the few abnormal cycles which may emerge throughout the day, doctors attach the device to the patient. The shapes of these waves and complexes are then interpreted by doctors. To establish whether the ECG displays symptoms of heart illness or not, they calculate parameters. The specifications include every wave's height and interval, which include the RR, PP, QT, and ST segments. For the doctor, identifying the fiducial points and computing the parameters is a tedious routine. To reduce the load of ECG interpretation, a system for automatic ECG recognition is desperately needed. For classifying cardiac arrhythmias, several studies have been conducted [1–4]. The MIT-BIH Arrhythmia Database [5] validates the suggested method. The researchers have developed a method for finding the QRS complex. A Discrete wavelet transform was used to segregate the ECG signal into time-frequency representations. The QRS complexes were initially developed in assembly language by Pan & Tompkins [6] for execution on a Z80 microprocessor, even though it requires less time and is easier to implement. This algorithm's main weakness is that QRS complexes' performance characteristics are adversely affected by frequency variation. Hamilton and Tompkins created a real-time QRS detection algorithm in the C programming language as a result [7]. The wavelet transform (WT) is being used by the researcher Cuiwei Li. The ECG waves and the QRS complex are simple to describe [8]. Many approaches to the problem of categorizing ECGs have been proposed in the literature, which include support vector machines [9], embedding EPR in medical imaging [10], genetic algorithm despeckling [11], healthcare and energy resources [12], noise removal [13], filter bank [14], artificial neural networks (ANNs) [15], self-organizing maps [16], and wavelet transform [17]. As a logical result, fuzzy sets have gained interest in present information technology [18–20]. To deal with uncertainty while making decisions in medical applications, machine learning (ML) techniques are crucial [21–23]. In this chapter, an innovative approach based on the WT with a particle swarm Optimization (PSO) technique to extract features and evaluate the ECG beats.

13.3 WAVELET TRANSFORM (WT)

ECG extracted features has been investigated since its inception, and many sophisticated algorithms and transformations have been developed for accurate and quick ECG feature extraction using the continuous WT. The effectiveness of an ECG signal is obtained by its fiducial points. WT is an adaptation of Short-Term Fourier Transform (STFT) that works on the principle of multi-resolution analysis of signal x[n] as shown in Figure 13.2.

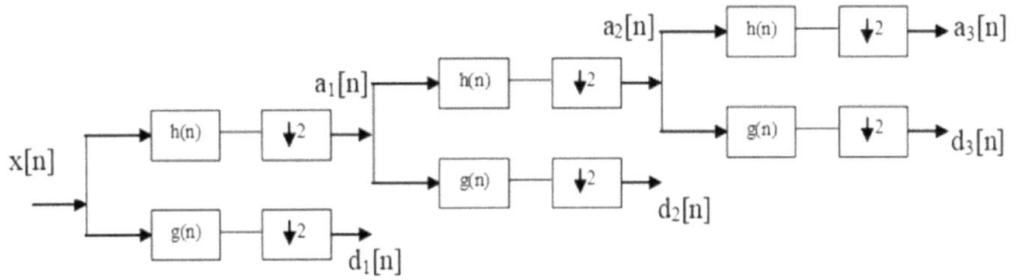

FIGURE 13.2 Three-level wavelet decomposition tree.

For all ECG decomposition phases, two filters and two down samplers are used. The filter's preparatory stage, g[n] functioning as a high pass filter denoted by d1, and h[n] operating as a low pass filter denoted by approximation a1. The accomplishment of several ECG decompositions is presented in the literature [28–30].

13.4 ECG DATABASE

In this chapter, used the PhysioBank database, which contains well-organized datasets of ECG signals and information related to biomedical research. The Atrial Arrhythmia and Malignant Ventricular Arrhythmia Databases are used to analyze these databascs [31].

13.5 PARTICLE SWARM OPTIMIZATION (PSO)

The PSO, the term "particles" refers to population members with a small mass and volume who are subjected to the fastest speeds or accelerations. Every particle in the swarm represents a solution in a high-dimensional space with four vectors: 1 current position; 2 after the particles' acceleration was compared to the old position, it selected the personal best position of all particles; 3 particle speed; and 4 the global best position for all particles that have yet to originate in its neighborhood. Every particle controls its position X_{ij} inside the restricted search region based on its own best position $XP_{best.ij}$ and the global best position $X_{Gbest.ij}$ acquired by all particles.

13.5.1 STEP FUNCTION

PSO is one of the bio-inspired algorithms and it is a simple one to look for a best solution in the solution space, which enhances the population randomization. Each individual particle's swarm will be sustained to represent the candidate solution by this method. The position in the optimal particle space represents the ideal answer to the problem, while the particle's velocity vector determines the particle's orientation and velocity value, each particle obeys the current optimal particle and finds the solution for optimization based on its own flying experience. Although the PSO offers several advantages, such as robustness, adaptability, and speed, it is easy for the MPP to slip into the local optimum during the tracking phase, resulting in a failure to reach the global peak. PSO in its traditional form uses the step function V_t^{ij}, which is computed for every particle ith of every jth variable at every iteration t, where V_t^{ij} 0, as shown in Eqn. (1). Particles are fixed by three mean the inertia function, which could be counted in every iteration from Eqn. (2); the inertia operator limit values and the third term are the maximum fit iteration number t_{max}; the second term refers to updating the current location's procedure, and the third term represents the method of sharing information of the global best solution between old locations. A key modification is that, as the acceleration factor of the convergence process, global optimization should also lead individual optimization to obtain the optimal solution, which may be represented mathematically using Eqn. (3). The other two alterations to the old speed function are the constant parameters c_1, c_2, c_3, and c_4, as well as the random numbers of MPSO, r_1, r_2, r_3, and r_4, which are between [0, 1].

$$V_{ij}^{t+1} = \omega^t v_{ij}^t + c_1 \gamma_1 \left(x_{Pbest \cdot ij}^t - x^t ij \right) + C_2 r_2 \left(x_{Gbest}^t - x_{ij}^t \right) \tag{1}$$

$$\omega^t = \omega_{max} - \left(\frac{\omega_{max} - \omega_{min}}{t_{max}} \right) \times t \, \forall \omega \geq 0 \tag{2}$$

$$V_{ij}^{t+1} = v_{ij}^{t+1} + C_3 R_3 (X_{Gbest.j}^t + X_{pbest.ij}^t - 2X_{ij}^t + C_4 R_4 (X_{Gbest.i}^t - X_{Pbest.ij}^t) \tag{3}$$

13.5.2 UPDATING OF ITH PARTICLE POSITION

Every particle should move closer to the ideal by updating its location with the step function V^t_{ij}, as shown in Eqn. (4), where the step is added to the current location. The step can be either positive or negative until it reaches zero, at which point it is the best option.

$$X_{ij}^{t+1} = X_{ij}^{t} + V_{ij}^{t+1} \quad (4)$$

13.5.3 SEARCH SPACE LIMIT REDUCTION

Consider narrowing the search region to find the best value, which is a new change to the search process that necessitates the adjustment and modification of the controlled variables' bounds in each iteration. As a result, the limit for the *Xi* controlled variable is reduced for it to converge to the global best location in every iteration when compared to its previous limit. Eqns. (5) and (6) could be used to calculate the limit reduction strategy (LRS) (Eqn. (6)).

$$X_{i.max}^{t+1} = X_{i.max}^{t} - \sigma(X_{i.max}^{t} - X_{Gbest.i}^{t}) \quad (5)$$

$$X_{i.min}^{t+1} = X_{i.min}^{t} - \sigma(X_{Gbest.i}^{t} - X_{i.\min}^{t}) \quad (6)$$

where; σ is a factor that is less than 0.1; depending on the problem, it is selected randomly and adjusted.

13.5.4 CONTROL OF VELOCITY LIMIT

The final change is that the velocity step or step length is limited based on the maximum/minimum values of the controlled variables X_i^{max}, X_i^{min} which is appropriate while approaching the obtained optimum solution. The velocity limitations V^i_{limits} can be adjusted to suit any case or system by changing the value of the velocity limit operator, as shown in Eqn. (7):

$$V_i^{limits} = \pm\beta(X_i^{max} - X_i^{min}) \quad (7)$$

The flowchart in Figure 13.3 depicts the entire implementation process of the proposed task utilizing MPSO. The duty ratio (*d*) of the converter is used as a particle's position, and the particle's fitness value is the generated PV power P_{pv} of the entire PV system.

If the output voltage of a DC-DC power converter is constant, the input voltage (PV output voltage) can be computed using the output voltage *Vo* and duty cycle (*d*). For example, the input voltage for the step-up converter utilized in this thesis can be computed as follows.

$$V_{in} = V_o \times (1-d) \quad (8)$$

As a result, the particle position (X^t_{ij}) in Eqns. (1)–(3) may be measured as the duty cycle of the PV converter, while the velocity (V^t_{ij}) can be measured as the variation of the duty cycle Δd^t_{ij} to run the PSO algorithm for PV systems. The best solution acquired by the particle itself (P_{best}) and the best solution in the entire population influence the variation of the duty cycle Δd^t_{ij} (G_{best}). If the current duty cycle d^t_{ij} cannot be determined using these two values, it will be updated at a high rate. P_{best} in Eqn. (9) will be updated if the condition in Eqn. (10) is satisfied; else, P_{best} will remain at its current value. The fitness value of each particle will then be calculated to see if the G_{best} value need updating.

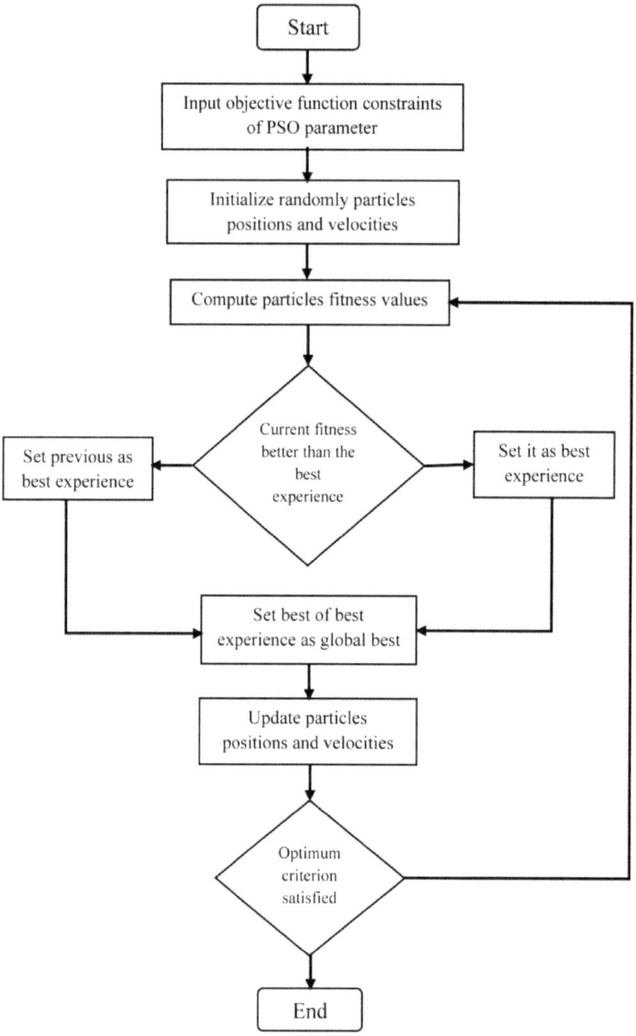

FIGURE 13.3 Particle swarm optimization process flowchart.

$$P_{best} = d_{ij}^{t} \tag{9}$$

$$f\left(d_{ij}^{t}\right) = f(P_{best}) \tag{10}$$

where; f is the goal function that needs to be improved.

13.6 METHODOLOGY

The multiresolution WT is the foundation of an ECG feature extraction system. The identification of R waves and the elimination of abnormalities in the ECG signal are important steps in establishing the purpose of this chapter. The main ECG signal features are identified in the feature extraction stage simply following the steps outlined below:

- **Step 1:** A band pass filtering is employed to the selected ECG record, which is among the MIT-BIH arrhythmia database records. All the data is a double-channel ECG signal. The WT, which would be utilized in ECG signal processing, splits the ECG signal into scales, making it much easier to analyze in different frequency ranges. The algorithm finds the absolute maxima of the discrete WT that exceed the specified threshold at each scale.
- **Step 2:** Several wavelet techniques are being used to perform four-level wavelet decomposition. The WT breaks down the ECG data into several frequency ranges, particularly zero crossings representing the ECG features waveforms. Depending on the WT, the filter is used to identify an R peak.
- **Step 3:** The QRS detection method involves identifying the point R of the heartbeat, which is usually the point with maximum amplitude. This detection also provides for the assessment of heart rate by measuring the distance between two adjacent QRS segments together within 150-ms timeframe. The algorithm examines the very first wavelet for both the peaks corresponding to the R-waves. The maximums and minimums are searched within a five-second search window to ensure that at least one peak relating to an R-wave is within search window.
- **Step 4:** The discrete WT computed using cubic spline wavelet at the scales a = 2^1, 2^2, and 2^3, respectively. The first wavelet is selected for R peaks which is greater than the adaptive threshold.

- **Step 5:** A fixed adaptive threshold value is greater than R waves and less than Premature Ventricular Contraction (PVC). PVCs are eliminated after they have been detected. The adaptive threshold methodology enables the selection thresholds to adapt to changes in signal quality, reducing the need for manual adjustments when changing from case to case. This chapter contains two types of adaptive threshold algorithm.

The first utilizes the first wavelet, with both the algorithm tracking the maximums and minimums that correspond to the QRS complexes. The second adaptive threshold algorithm, like first, explores for maximums and minimums and determines the wavelet amplitude of normal R waves. Additionally, wavelet filtering is often used in ECG registration, using Haar and Db4 wavelets. Figure 13.4 shows how well R waves are identified in various information. Determining R wave latency is extremely difficult. Wavelet transformations of some ECG signals were not performed accurately due to very high frequency noise, baseline drift, and artifacts. The presence of high amplitude peaks rendered recognizing R-waves in some signals incredibly hard.

The algorithm is now applied on 35 records of MIT-BIH using cubic spline wavelet. This Database to show the efficiency of Symlet Wavelet. The QRS detection algorithm was determined by using the Percentage of Sensitivity (Se) and Positive Predictivity (PP) are given by Eqns. (11) and (12) listed below.

The sensitivity defines the proportion of the beats that had been suitably identified by the algorithm.

$$\text{Sensitivity}\left(\%\right) = \frac{TP}{TP + FN} \quad (11)$$

Positive predictivity is defined as the ratio of number of correctly classify the normal heart beats to total number of heart beats.

$$\text{Predictivity}(\%) = \frac{TP}{TP + FP} \quad (12)$$

where; true positive (TP) is the correctly detected; false negative (FN) is the number of missed detections; and false positive (FP) is the excess of heartbeat detection in QRS detection.

Table 13.1. The find the R peak detection using MIT-BIH Database to get the result of sensitivity is 99.72% and its Positive Predictivity is 99.85%.

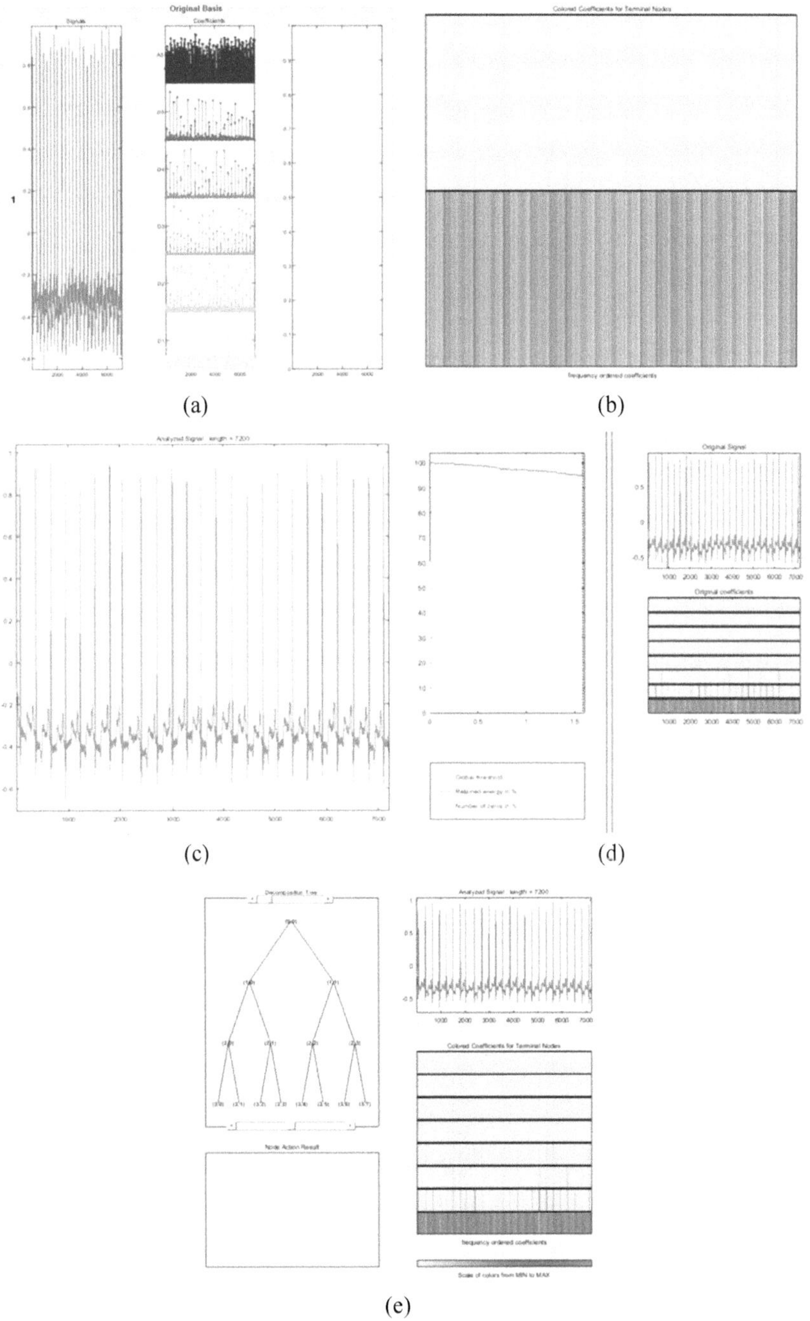

(a) (b) (c) (d) (e)

FIGURE 13.4 (a) Original ECG signal; (b) co-efficient of terminated nodes; (c) detection of QRS complex; (d) fixation of threshold; and (e) decomposition tree.

TABLE 13.1 The Results of R Peak Detection Using Symlet Spline Wavelet Transform

Number of ECG Record	TP	FN	FP	Sensitivity (%)	Positive Predictivity (%)
100	2,261	11	0	99.53	100
104	2,215	15	48	99.45	97.97
105	2,543	11	4	99.57	99.88
106	2,012	16	6	99.25	99.65
108	1,768	7	1	99.60	99.94
109	2,526	4	3	99.84	99.88
111	2,126	1	0	99.95	100.00
112	2,538	0	0	100.00	100.00
113	1,794	0	2	100.00	99.93
114	1,878	1	2	99.95	99.94
115	1,941	5	1	99.74	99.83
121	1,863	4	1	99.78	99.94
122	2,476	0	1	100.00	99.95
202	2,095	7	1	99.66	99.95
203	2,963	14	65	99.52	99.62
205	2,256	17	1	99.36	99.96
207	2,950	86	121	96.31	94.89
208	2,955	31	12	98.96	99.59
209	3,004	0	3	100.00	99.90
210	2,650	23	11	99.13	99.58
212	2,647	2	0	99.92	100.00
213	3,251	7	2	99.78	99.93
214	2,261	6	0	99.67	100.00
215	3,248	4	0	99.87	100.00
217	2,208	3	1	99.86	99.95
219	2,154	4	1	99.81	99.95
220	2,047	0	2	100.00	99.90
222	2,483	0	7	100.00	99.71
223	2,605	3	0	99.88	100.00
228	2,053	7	26	99.66	98.74
230	2,256	0	0	100.00	100.00
231	1,569	0	1	100.00	99.93
132	1,779	0	2	100.00	99.88
233	3,075	14	10	99.87	99.67
234	2,750	3	1	99.89	99.96
Total	**83,200**	**306**	**336**	**99.72**	**99.85**

13.7 RESULTS AND DISCUSSIONS

In this chapter, we examined 35 ECG datasets from the MIT-BIH Database, which comprise of five distinct forms of arrhythmias such as Normal Sinus Rhythm (NSR), Myocardial Infraction (MI), Ventricular Tachycardia (VT), Ventricular Fibrillation (VF), and Ventricular Flutter (VFLU). Using the cubic spline WT, the Energy level may be computed. As indicated in Table 13.2, different sets of ECG waveforms were formed into five separate classes and five output layers. Each group's output vector is determined by the output target vector in Boolean value. Each entry in the MIT-BIH Database has an annotation file that expert physicians can use to identify it. This annotated data can be used to design the target vector and assess classifier performance.

$$\text{Classification Accuracy} = \frac{\text{Number of beats correctly classified}}{\text{Total number of beats}} \quad (13)$$

TABLE 13.2 Classification Accuracy of PSO

Cardiac Abnormalities	Training of Beats	Testing of Beats	Undetected Beats	Accuracy (%)
NSR	105	45	1	99.25
MI	85	52	0	100.00
VT	60	30	1	99.33
VF	96	47	0	100.00
VFLU	90	65	1	99.12
Total	**436**	**239**	**3**	**99.5**

13.8 CONCLUSION

This chapter has presented to eliminate the noise in ECG and easy to identify the QRS complex detection based on cubic spline WT. Because it reduces the probability of inaccuracy, the proposed simulation is best suited for this clinical application. The extracted ECG features are used to recognize and monitor cardiac abnormalities automatically. The studies were carried out for 5 distinct heartbeat types from the MIT-BIH database. The overall classification accuracy of the DNN system is 99.5%. In the future, the ECG distinctive points would be recovered employing the singularities detection approach and the multi-WT. The simulation results are highly reliable, and the classifier could provide more help to identify cardiac conditions at each stage.

KEYWORDS

- **artificial neural networks**
- **cardiovascular disease**
- **particle swarm optimization**
- **short-term Fourier transform**
- **wavelet transform**
- **World Health Organization**

REFERENCES

1. Chen, (2000). A two-stage discrimination of cardiac arrhythmias using a total least squares-based Prony modeling algorithm. *IEEE Trans on BME., 47*(10), 1317–1327.
2. Inan, G., & Elif, D. U., (2005). ECG beat classifier designed by combined neural network model. *Pattern Recognition, 38*, 199–298.
3. Yasushi, K., & Koji, O., (2004). A study for excluding incorrect detections of Holter ECG data using SVM. In: *2004 ICONIP* (pp. 1223–1228).
4. Stanislaw, O., Linh, T. H., & Tomasz, M., (2004). Support vector machine-based expert system for reliable heartbeat recognition. *IEEE Transactions on Biomedical Engineering, 51*, 582–589.
5. MIT-BIH. http://www.physionet.org (accessed on 20 November 2023).
6. Pan, J., & Tompkins, W. J., (1985). A real-time QRS detection algorithm. *IEEE Transaction Biomedical Engineering, 32*, 230–236.
7. Hamilton, P. S., & Tompkins, W. J., (1986). Quantitative investigation of QRS detection rules using the MIT/BIH arrhythmia database. *IEEE Transactions on Biomedical Engineering, 33*, 1157–1165.
8. Cuiwei, L., Zheng, C., & Tai, C., (1995). Detection of ECG characteristic points using wavelet transform. *IEEE Transactions on Biomedical Engineering, 42*, 8–21.
9. Qibin, Z., & Liqing, Z., (2005). ECG feature extraction and classification using wavelet transform and support vector machines. *IEEE Transactions on Biomedical Engineering., 23*, 1089–1092.
10. Sonal, A., Lakshmi, C., & Varun, M., (2020). A secure reversible data hiding and encryption system for embedding EPR in medical images. *Current Signal Transduction Therapy, 15*, 124–135.
11. Mokeddem, D., & Khellaf, A., (2007). Design of a fuzzy controller for pH using genetic algorithm. *International Review of Computer Science and Software, 2*, 318–324.
12. Davazdah-Emami, H., Emad, R., & Mohammad, N., (2019). Considering environmental health and energy resources to design transformers. *Current Signal Transduction Therapy, 14*, 284–293.

13. Hari, K. P., (2019). Speckle noise reduction using ant colony-based Gabor filter methodology for improved fetal parameter estimation of US images. *Current Signal Transduction Therapy, 14*, 225–232.
14. Afonso, X., Tompkins, W. J., Nguyen, T. Q., & Luo, S., (1999). ECG beat detection using filter banks. *IEEE Transactions on Biomedical Engineering, 46*, 192–202.
15. Louis, C. P., & Cobus, N., (2002). Feature extraction from ECG for classification by artificial neural network. University of Pretoria. In: *IEEE Proceeding* (pp. 67–70).
16. Talbi, M. L., & Charef, A., (2009). PVC discrimination using the QRS power spectrum and self-organizing maps. *Computer Methods and Programs in Medicine, 22*, 45–52.
17. Mahmoodabadi, S. Z., Ahmadian, A., Abolhasani, M. D., Eslami, M., & Bidgoli, J. H., (2005). ECG feature extraction based on multiresolution wavelet transform. *IEEE EMBS, 27th Annual International Conference* (pp. 1142–1144).
18. Anuradha, B., Suresh, K. K., & Veera, R. V. C., (2008). Cardiac arrhythmia classification fuzzy classifiers. *Journal of Theoretical and Applied Information Technology, 4*, 353–358.
19. Benali, R., & Chikh, M. A., (2009). Neuro-fuzzy classifier for cardiac arrhythmias recognition. *Journal of Theoretical and Applied Information Technology, 4*, 577–582.
20. Kannathal, N., Lim, C. M., Acharya, U. R., & Sadasivan, P. K., (2006). Cardiac state diagnosis using adaptive neuro-fuzzy technique. *Medical Engineering and Physics, 28*(8), 809–815.
21. Emran, M. T., & Nor, H. K., (2009). Heartbeat electrocardiogram signal feature extraction using discrete wavelet transform. *Proceedings of CSPA*, 1112–1117.
22. Daubechies, I., (1990). The wavelet transforms, time-frequency localization and signal analysis. *IEEE Transaction and Information Theory*, 961–1005.
23. Sumathi, S., & Sanavullah, M. Y., (2009). Comparative study of QRS complex detection in ECG based on discrete wavelet transform. *International Journal of Recent Trends in Engineering, 5*, 273–277.

CHAPTER 14

ROBOTS AND CYBORGS WITH ARTIFICIAL INTELLIGENCE-BASED TECHNOLOGIES IN THE HEALTHCARE SECTOR: A REVIEW

S. PONLATHA[1] and S. SUMATHI[2]

[1]*Department of Electronics and Communication Engineering, Mahendra Engineering College, Tamil Nadu, India*

[2]*Department of Electrical and Electronics Engineering, Mahendra Engineering College, Tamil Nadu, India*

ABSTRACT

The fast advancement of robotic technology opens up new avenues for biomedical and healthcare engineering. The new model is created, and it is used to determine the selection criteria among robot, cyborg, and human services, with an emphasis on the healthcare service sector. This research recommends the employment of robots and cyborgs as surgeons in eye surgery in the future. Following that, the established model was used to investigate the intention to utilize each surgeon (i.e., robot surgeon, cyborg surgeon, and human surgeon). The PLS-SEM method was used to gather and evaluate the data. The research findings revealed that social impact, perceived danger, work expectation, and behavioral intention all had a substantial impact on respondents' intentions to utilize robot services. Nevertheless, the cyborg service model's findings supported the notion that the desire to utilize cyborg operations is strongly impacted by the expectation on commitment, arousal, achievement, and cultural pressure. Further evidence of their massive

Robotics and Automation in Healthcare: Advanced Applications. R. Thanigaivelan, Sanjay Singh, & Clement Christy Deepak C. (Eds.)

influence on the willingness to use social work came from task expectation and social influence. The outcomes of the three models also revealed that, with varying degrees of significance depending on the model for effort expectancy, social influence and behavioral intention strongly influenced the intention to employ these surgical services. The ANOVA t-test supported the differences in the respondents' intentions to select one of the suggested surgeons, which were supported by their preference for a human surgeon over a cyborg and a robot, respectively.

14.1 INTRODUCTION

A perception of tech's ability to improve innate human capacities, combined with cultural assumptions about what is considered incomplete, normal, or improved, leads to a variety of body-altering techniques that may not only restore functions, but may also go beyond what is typically considered therapeutic intervention. Accordingly, a strong argument exists regarding the ability to enhance humans' body according to particular needs or desires [1]. In this context, the enhancements are related to upgrading human capabilities, which opens a new type of research concerning superhuman or what is called "Cyborg" [2]. Cyborg can be defined as a cognitively or bodily enhancement of human capabilities, and it can be called a transhumanism. In general, transhumanism and technology convergence are directing the scientist's efforts toward enhancing human social skills, health, happiness, intelligence, and performance. In effect, cyborg technology has different forms, such as technological implants, brain-computer interfaces (BCI), externalization of cognitive functions, and neuronal prosthesis [3].

Greguric [21] pointed to four research areas affecting the development of human enhancement technologies: cognitive sciences, nanotechnologies, information technologies, and biotechnologies. Furthermore, the enhancements can be categorized into:

1. **Cognitive Abilities Enhancement:** Enhancement of cognitive functions include improved memory, decision-making, and sensory processing. These abilities could be achieved by using technological implants or wearables technology.
2. **Physical Capabilities Enhancement:** Improvement of one's physical attributes, such as power, endurance, and precision. These enhancements could be created by using bionic technology, genetic engineering, wearables technology, and pharmacology.

14.2 CYBORG TECHNOLOGIES

14.2.1 COCHLEAR IMPLANT (CI)

The cochlear implant (CI), the first widely utilized implant technology, is used to restore hearing abilities by replacing the hearing cells and converting sounds into electrical impulses that stimulate the auditory nerve's function [4]. From a medical perspective, is the CI prosthesis? It could be considered as prosthesis, addition, replacement, extension, augmentation, or even enhancement. Nowadays, the term "cyborg" is used to compensate for the term "prosthesis" in academic research and discussion forums about implants. Also, human engagement with the world is surely changed by the use of CI, and some CI users are introducing themselves as biotic or cyborg entities [5, 6]. As per Lee [7], CI considered as a hearing aid, and it may become a part of human bodily functions, because the future may turn the use of it from therapy toward enhancement, to increase human hearing capabilities to become beyond the normality.

14.2.2 NEUROMOTOR PROSTHESES (NMPS)

The purpose of using NMPs is to restore lost motor functions in paralyzed humans, through transferring signals related to the motion from the brain and around the damaged parts of the nervous system into external effectors (Figure 14.1). It works by using neural probe arrays implanted into the motor areas of the cerebral cortex for recording the activity of neurons, then computer algorithms translate the pattern of neural firing rates to produce output control signals to the external device, such as robotic arm, and depending on the user thoughts [8].

14.2.3 DEEP BRAIN STIMULATION (DBS)

DBS is the implantation of a neurostimulator, a medical device that delivers electrical impulses to a certain targeted area of the brain, using two implanted electrodes, to treat movement and neuropsychiatric problems (Figure 14.2). Treatment-resistant conditions include Parkinson's disease, essential tremor, dystonia, chronic pain, significant depression, and obsessive-compulsive disorder can benefit therapeutically from this technique. Although its efficacy in treating a few illnesses has been shown, there are still possible adverse effects and grave consequences [9].

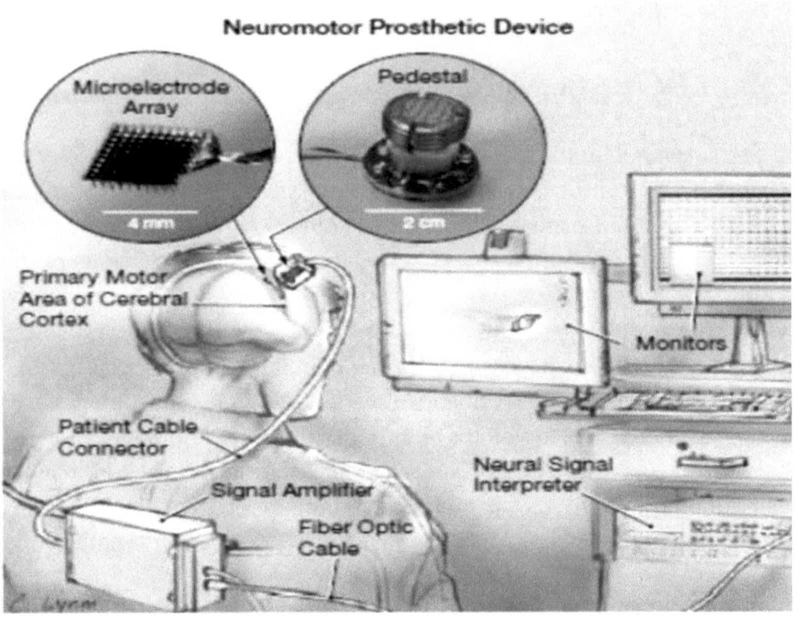

FIGURE 14.1 Neuro motor prostheses.

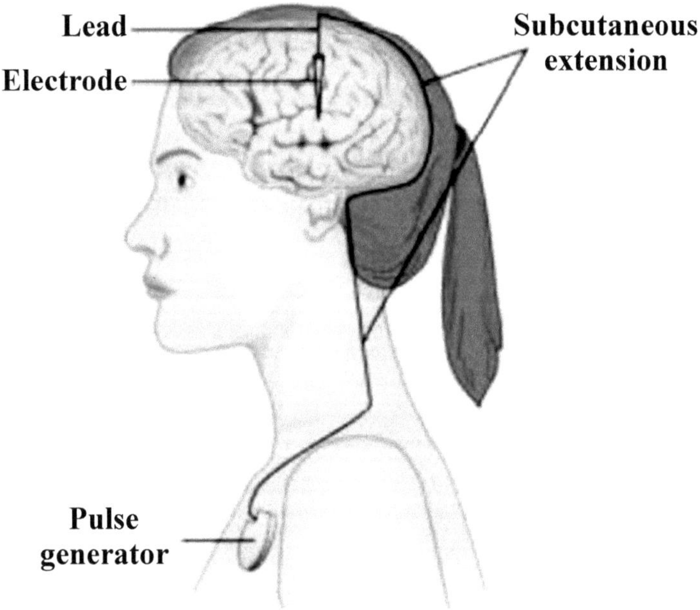

FIGURE 14.2 Deep brain stimulation process.

14.2.4 NEUROFEEDBACK

Neuro feedback is another type of human-machine interaction, where the brain activities are detected by devices like electroencephalography (EEG) through placing sensors in the scalp to measure these activities. For instance, this principle is used for preventing epileptic attacks with the aid of Vagus Nerve Stimulation (VNS) [10]. The VNS is a device implanted in the human body to stimulates the Vagus nerve by using electrical impulses. In reality, there is one Vagus nerve in each body's sides, running from the brainstem through the neck to the chest and abdomen. When alternative therapies fail to control seizures, VNS technology is used. Also, it can be used for hard-to-treat depression that hasn't responded to typical therapies. Researchers are trying to use VNS in the treatment of multiple sclerosis, headache, pain, and Alzheimer's disease (Figure 14.3).

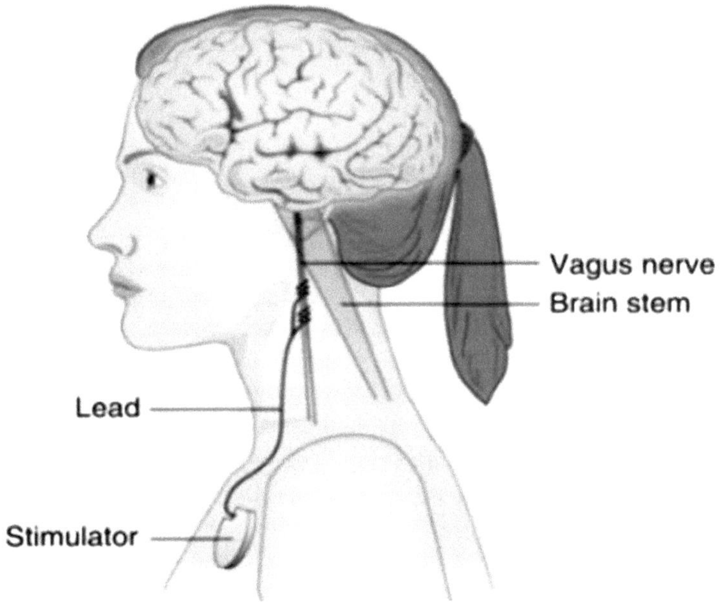

FIGURE 14.3 Implementation of neurofeedback.

14.2.5 ALL-CHAIN-WIRELESS BRAIN-TO-BRAIN SYSTEM

All-chain-wireless Brain-To-Brain System (BTBS) that can control the motion of cyborg cockroach through the human brain. A portable microstimulator was integrated surgically into cyborg cockroach to produce invasive stimulation of an electrical nerve. A BCI with Steady-State Visual Evoked Potential (SSVEP) was used to recognize human motion commands, and an optimization algorithm was suggested in SSVEP to improve the online performance of the BCI. A specific train of electrical impulses was triggered by microstimulator using BCI commands and sent through antenna nerve to stimulate the brain of cockroach via Bluetooth technology. The proposed BTBS enabled a practical and functional pathway to transfer the information from the human brain to the cockroach brain.

14.2.6 RETINAL STIMULATION

Retinal stimulation acts as a realistic solution for restoring vision to the blind people. Recently, different clinical experiments showed the ability of visual prostheses, especially retinal implants, to restore vision. A new concept of neuronal stimulation besides the traditional electrical stimulation, which is based on implanting metal electrodes in different areas of the visual path, is used to build the next generation of retinal implants. But vision restoring technology is still under investigation and needs further cooperation between neuroscientists, material science, clinicians, and micro-fabrication sciences.

14.3 OPERATIONAL AND STRUCTURAL PROPERTIES OF A CYBORG

A silicon chip is implanted into any portion of the body, notably the region where most of the nerves are attached and send and receive the electrical impulses. This silicon chip is developed in such a manner that it can receive the nerve impulses, amplify them, and encode the signal into digital format by which suitable computer accessibility is supplied. Since no wires are preferred to interface the cyborg and computer, a wireless communication channel is recommended. This implant is housed in a glass tube. One includes the power supply, a copper coil activated with regard to the signals from the Cyborg. Radio waves to create an electric current. In the other end, three small, printed circuit boards will transmit and receive signals. The implant is connected to the body by a band that wraps around the nerve fibers and is attached by a very tiny wire to the glass capsule. The chips in the implant will receive signals from the nerve fibers and relay them to a computer immediately. For example, when a finger is moved, an electrical signal passes from the brain to trigger the muscles and tendons that function the hand. The finger will continue to receive these nerve impulses. The signal from the implant will be analog, thus it needs to be transformed into digital in order to save it in the computer. Still, various research on labor, organizational culture, computerized information systems (CIS), networks, and human-machine dyads (such as the "symbolic value of the CIS" or the "organizational symbolism" of computer Culture) show the urge to study.

14.4 IMPLANTATION PROCEDURE

Not only may this technology benefit blind individuals, but it may also be feasible to record impulses responsible for happiness, pain, anesthetic, and other emotions. Experiments are also being carried out to create wireless communication between two people by implanting similar chips in their bodies that can use energy to transmit and receive impulse impulses. If this experiment is successful, cyborgs, which were previously thought to be science fiction, will no longer be a faraway fantasy. This approach does not require the use of words to communicate; instead, impulses in the human body may be used to send information between each other. Telephones will be rendered obsolete by thought communication. Another significant application of cyborgs would be in illness treatment. If this sort of experiment is successful, we can expect researchers to discover how to transmit antidepressant stimulation, contraceptives, or immunizations in a similar fashion. This has the potential to change the face of medicine, to forsake the paradigm of giving people chemical treatments and cures in favor of achieving the desired outcomes electronically. Cyber medicines and cyber narcotics have the potential to heal cancer and alleviate severe depression.

14.5 ROBOT TECHNOLOGIES

Before a century, robots' ability for autonomous mobility and to perform a set of tasks had been captured by writers' imaginations. While recently, robots have emerged from the pages of science fiction novels into the real world [11]. Different types of robots were developed in the past few decades, such as industrial, mobile, educational, collaborative, and service robots [12]. Industrial robot (Figure 14.4) is defined as a reprogrammable and automatically controlled robot that has the ability to perform a multipurpose manipulation with three or more programmable axes. It could be fixed or mobile and used in industrial automation applications [22].

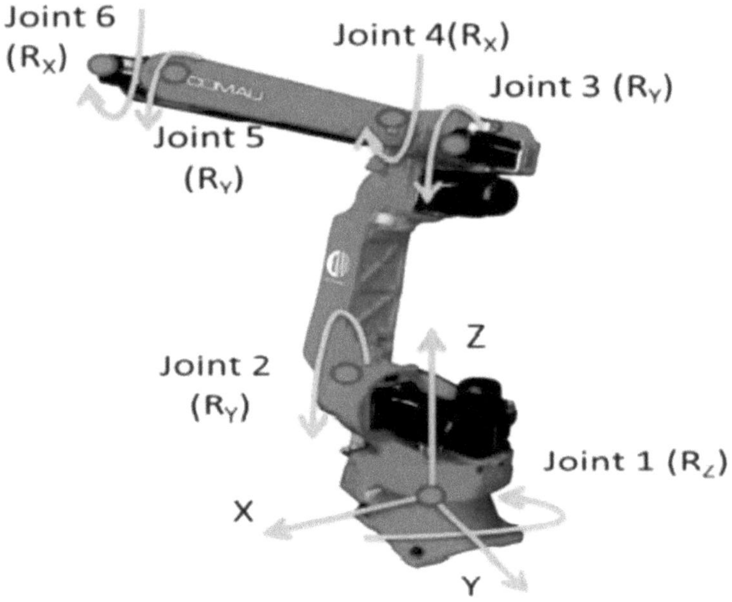

FIGURE 14.4 Industrial robot.

Service robots are set of mobile robots, designed to work in populated environments, such as hospitals, offices, restaurants, universities, museums, and homes. They are developed to perform different tasks, like cleaning, education, entertainment, and eldercare. Likewise, there are some of the autonomous and intelligent robots that are in use for home assistant tasks.

14.6 ARTIFICIAL INTELLIGENCE (AI) AND MACHINE LEARNING (ML) TECHNOLOGIES

AI and ML technologies are embedded in the robots' technology, and the vast development in both of them enables new features and improvements in robot design. For example, motion control, vision, grasping, and understanding data patterns are some of AI and ML inventions, which are implemented in robots. AI is considered as a study of human intelligence behaviors (e.g., recognition, problem-solving, learning, perception, language use, creativity, and manipulation of symbols) to produce models that can be used in creating computer programs that have the ability to simulate human intelligence. These programs can be used for different purposes, such as voice recognition, shape recognition, theorem proving, games running, language translation, music composition, medical diagnosis formation, and expertise formation.

Recently, the development of ML is driven by new learning algorithms, big-data, and inexpensive computation systems. For example, in the healthcare sector, ML is considered a flexible and powerful instrument that can be used to resolve and predict results from clinical and biological information. Its models have the power to improve the efficiency of healthcare in different ways, such as in implementing diagnostic models that can be used in risk stratification, in the medical evaluation and suggested treatments [13]. Furthermore, various ML methods have been successfully utilized in emotional recognition to predict individuals' emotional state, by using a set of physiological features as a reference. In this context, the robots need to be able to capture the companion's emotional state. Realizing emotional status from physiological cues is an efficient method for achieving successful human-robot interaction.

Robot learning is a result of the interaction between ML and robot technology. It is related to:

1. **Reinforcement Learning:** It is a type of ML that is suitable for use with robots. It assumes that the world consists of various contexts and agents (i.e., robots) states S, the agent can execute a set of actions A, roles to explain what the agent observes, and the time which is represented by discrete steps. The agent will be given a reward for each performed action as shown in Figure 14.5.

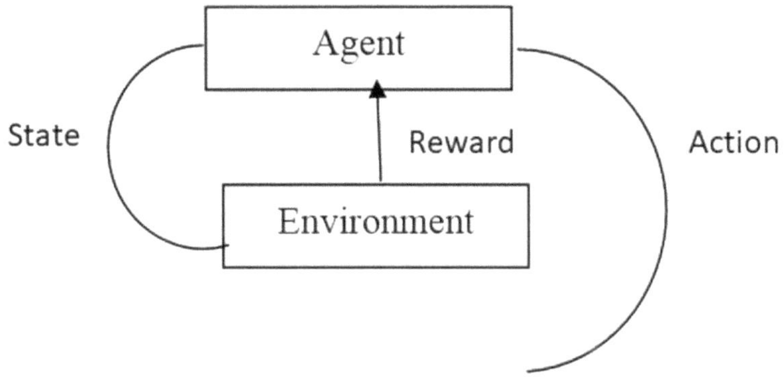

FIGURE 14.5 Reinforcement learning framework.

The agent (Figure 14.6) has the following components [14]:

i. **Sensors:** to collect data on the condition of the immediate surroundings.
ii. **Effectors:** to provide the agent the ability to alter its surroundings.
iii. **Reinforcement Mechanism:** A signal is generated once each action is completed.
iv. **Reinforcement Learning:** it is related to the feedback information from the environment.
v. **Control Policy:** it is related to situation-action planning.

The reinforcement learning agent is reactive, which means it can transform the collected experience into situation-action rules (Control Policy), and adaptive that uses feedback information to enhance its performance. Also, when compared to other learning systems, it is less dependent on trainers because it is learning from reinforcement mechanism, and less dependent on the previous field of knowledge since it's an inductive learner. Moreover, the reinforcement learning has different types of algorithms for a different use, such as Q-learning, deep Q network (DQN), asynchronous actor-critic algorithm (A3C), and deep deterministic policy gradient (DPG). For instance, DQM, which applied in the Atari 2600 video games, combines reinforcement learning with ANNs.

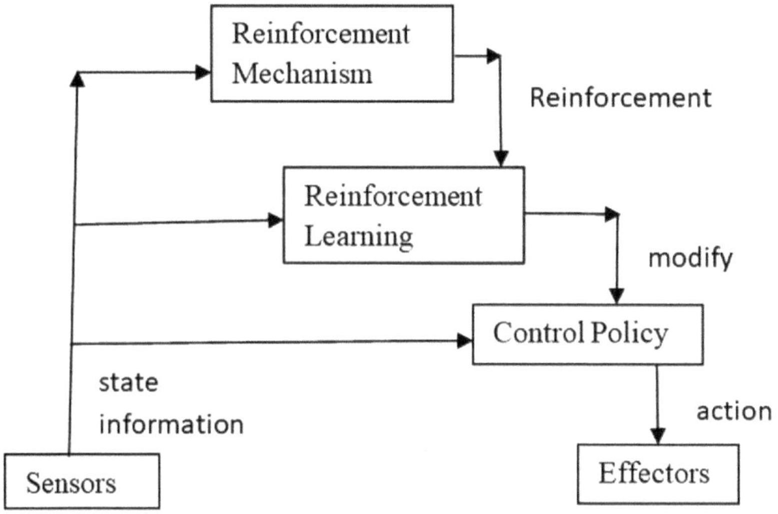

FIGURE 14.6 Reinforcement learning paradigm.

2. **Developmental Robotics:** It concerns in studying the development of lifelong cognitive behaviors and it focuses on robot Autonomous Mental Development (AMD), such as artificial emotions, self-organization, and self-motivation [15]. AMD is related to the use of robot sensors and effectors to develop the mental capabilities by the autonomous interaction with its environment.
3. **Evolutionary Robotics (ER):** It is the application of an artificial peer of natural Darwinian evolution to develop robots' artificial brains, sensors, and bodies [16]. It based on evolutionary computation by repeating cycles of controller convenience and selection, which are almost similar to the natural evolution of generations. For each cycle, each controller from a population of candidate controllers is set to perform a specific task, and to be involved after that in an evaluation period. The evaluation is based on a fitness function, which is a function used to measure if the design is fulfilling its targets, and modification is repeatedly performed based on the fitness function evaluation. Then, a Genetic Algorithm (GA) is applied based on the information resulting from the fitness function to select and replicate the fittest candidates for next generation.

14.7 RESEARCH MODEL

Figure 14.7 represents the theoretical model that explains the intention to use among humans, robots, and cyborgs services, which has been defined based on the aforementioned hypotheses.

- **H1a:** Consumer intention to use among human, robot, and cyborg services is affected positively by Effort Expectancy.
- **H1b:** Consumer intention to use among human, robot, and cyborg services is affected positively by Performance Expectancy.
- **H2:** Consumer intention to use among human, robot, and cyborg services is positively affected by the favorable Social Influence.
- **H3a:** Consumer intention to use among human, robot, and cyborg services is positively affected by Pleasure.
- **H3b:** Consumer intention to use among human, robot, and cyborg services is positively affected by Arousal.
- **H4:** Consumer intention to use among human, robot, and cyborg services is affected negatively by the Perceived Risk.
- **H5:** Consumer intention to use among human, robot, and cyborg services is affected positively by Perceived Empathy.

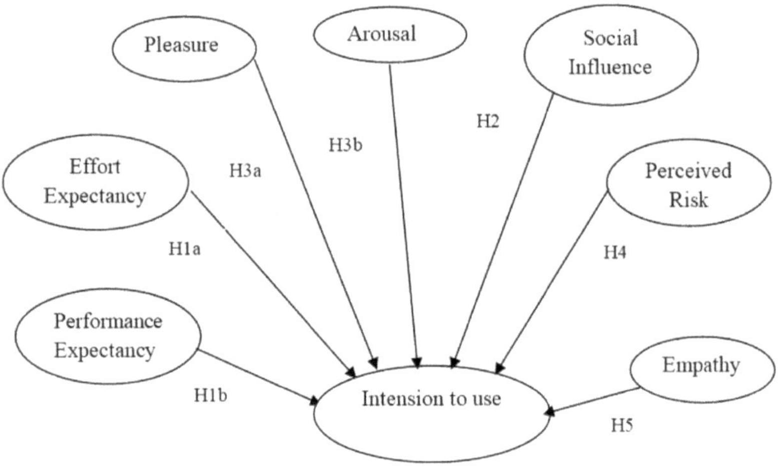

FIGURE 14.7 Theoretical model of acceptance of robot, cyborg, and human services.

14.8 DATA ANALYSIS PROCESS

Structural equation modeling (SEM) has gained popularity in recent years, particularly in social science research [17]. SEM is related with two methods: Covariance-Based SEM (CB-SEM) and Partial Least Squares SEM (PLS-SEM). In general, CB-SEM is most commonly employed in confirmatory research, where the goal is to confirm or reject ideas. These ideas are represented by a collection of variables with a systematic connection that may be empirically evaluated. This research model was examined using the partial least-square structural equation modeling (PLS-SEM) approach. This technique employs a component-based approach and allows for the assessment of both the measurement model and the structural model at the same time. Furthermore, it is an appropriate strategy for dealing with several relationships at once, does not require a normal distribution, and is suggested for examining complicated frameworks. The PLS-SEM, on the other hand, is more likely to be utilized in exploratory research (i.e., predictive applications and theory building) by evaluating theoretical models to explain the variation of the dependent variable. In order to maximize the explained variance of the endogenous latent variables, PLS-SEM evaluates the model connections in a series of ordinary least squares (OLS) regressions. The series of OLS regressions allows PLS-SEM to reach a better degree of statistical power while using less sample size. CB-SEM, on the other hand, reproduces an experimentally observed covariance matrix by estimating model parameters.

When deciding between the two approaches, the following criteria may be relevant:

- Select PLS-SEM if the main objective is to predict or identify the key constructs.
- Select PLS-SEM if the main objective is to extend an existing theory or if the research is exploratory.
- Select the CB-SEM if the main objective is to test a theory, confirm a theory, or compare alternative theories.

14.9 RESULTS AND DISCUSSIONS

The information was gathered from 379 people from various universities in South India. The internal consistency reliability is the first criterion to be evaluated by using Cronbach's alpha and composite reliability. This research results confirmed the internal consistency reliability. The values of Cronbach's alpha and the composite reliability for all the robot services measurement model were higher than 0.70. Furthermore, the standardized loading of the indicators should be greater than 0.70 and the t-value should be greater than 1.96 to ensure the correct reliability indicator in the measurement model. Even though the values between 0.4 and 0.7 could be accepted if the t-value is above 1.96.

The positive correlation between the indicator and the alternative indicators of the same constructs means that the convergent validity is met. To obtain this, the Average Variance Extracted (AVE) value should be equal to or greater than 0.50. In this research, all constructs of the robot services measurement model had AVE values greater than 0.50, which confirmed the convergent validity of the measurement model. Table 14.1 shows loading values, Cronbach's alpha, Composite reliability, and AVE values. Measuring the discriminant validity is the last step in the evaluation of the measurement model.

To examine discriminant validity, two approaches were utilized. The first is the Fornell-Larcker criteria, which compares the latent variable's correlation with the square root of the AVE values. To achieve the discriminant validity requirement, the square root of the AVE value for each construct must be larger than its correlation with the other components. The second technique involves calculating the relationships' Heterotrait-Monotrait ratio (HTMT). The findings of the robot services measurement model validated the discriminant validity of the construct indicators, as the HTMT values of all constructs were less than 0.90, and the square root of AVE values for each construct was more than the correlation value with the other constructs. Table 14.2 displays the HTMT, AVE square root, and correlation values for everyone.

TABLE 14.1 Internal Consistency Reliability and Convergent Validity for Robot Services Measurement Model

Variable	Indicators	Loading	Cronbach's Alpha	Composite Reliability	AVE
Arousal	RA1	0.945	0.875	0.941	0.899
	RA2	0.940			
Effort expectancy	REE1	0.929	0.953	0.966	0.875
	REE2	0.934			
	REE3	0.939			
	REE4	0.940			
Empathy	RE1	0.849	0.936	0.952	0.798
	RE2	0.841			
	RE3	0.935			
	RE4	0.925			
	RE5	0.913			
Intension to use	RIU1	0.963	0.926	0.964	0.931
	RIU2	0.966			
Pleasure	RP1	0.937	0.822	0.918	0.848
	RP2	0.904			
Perceived risk	RPR1	0.893	0.873	0.917	0.786
	RPR2	0.823			
	RPR3	0.940			
Performance expectancy	RPE1	0.942	0.941	0.957	0.849
	RPE2	0.925			
	RPE3	0.932			
	RPE4	0.886			
Social influence	RSI1	0.932	0.926	0.953	0.871
	RSI2	0.944			
	RSI3	0.923			

14.10 CONCLUSION

The robotic technology is primarily available in the market for various applications; however, it is also being developed for enhanced functions in areas such as healthcare services, with the primary goal of using this technology to replicate a surgeon's tasks and perform autonomously without

TABLE 14.2 Discriminant Validity of the Robot Services Measurement Model

Variable	Arousal	Effort Expectancy	Empathy	Intention to Use	Perceived Risk	Performance Expectancy	Pleasure	Social Influence
Arousal	0.943	0.619	0.602	0.698	0.100	0.722	0.856	0.706
Effort expectancy	0.566	0.935	0.711	0.795	0.049	0.777	0.614	0.774
Empathy	0.546	0.673	0.893	0.543	0.240	0.592	0.601	0.663
Intention to use	0.629	0.748	0.507	0.965	0.109	0.860	0.710	0.815
Perceived risk	–0.098	0.040	0.229	–0.111	0.887	0.055	0.058	0.079
Performance expectancy	0.655	0.738	0.557	0.805	–0.002	0.921	0.721	0.881
Pleasure	0.728	0.548	0.533	0.625	–0.045	0.641	0.921	0.690
Social influence	0.636	0.728	0.618	0.757	0.056	0.824	0.609	0.933

Note: The values on the main diagonal are the square roots of the AVEs, below the diagonal: correlations between the constructs, and above the diagonal: HTMT values.

human intervention. The findings of the three models revealed that social influence and effort expectation had a considerable impact on the intention to use these surgical services, with the intensity varying between models for effort expectancy. The impact of social influence provides a broad overview of the nature of the healthcare industry. The effort expectancy impact influences patients' expectations of ease of use and contact with the recommended surgeons. According to PLS-SEM findings, the robot services model described 73.8% of the variation in the intention to employ robot healthcare services, the cyborg services model explained 77.0%, and the human services model explained 77.7%. This suggests that all of the models developed in this research are highly predictive of the desire to utilize the relevant services.

The models assessed performance expectancy, effort expectancy, social influence, empathy, perceived risk, and emotions (Pleasure and arousal) variables. Four of the examined variables affected the intention to use robot service, except empathy and emotional dimensions.

KEYWORDS

- **brain-computer interfaces**
- **brain-to-brain system**
- **cochlear implant**
- **deep brain stimulation**
- **electroencephalography**
- **neuromotor prostheses**
- **Vagus nerve stimulation**

REFERENCES

1. Hogle, L. F., (2005). Enhancement technologies and the body. *Annual Review of Anthropology, 34*, 695–716. https://doi.org/10.1146/annurev.anthro.33.070203.144020.
2. Parkhurst, A., (2012). Becoming cyborgian: Procrastinating the singularity. *The New Bioethics, 18*(1), 68–80. https://doi.org/10.1179/2050287713Z.0000000006.
3. Romportl, J., (2015). *Beyond Artificial Intelligence: The Disappearing Human-Machine Divide*. https://doi.org/10.1007/978-3-319-09668-1.

4. Jarrett, C., (2013). The age of the superhuman. *British Psychological Society, 26*(10), 720–724.
5. Christie, E., & Bloustien, G., (2010). I-cyborg: Disability, affect, and public pedagogy. *Discourse: Studies in the Cultural Politics of Education, 31*(4), 483–498. https://doi.org/10.1080/01596306.2010.504364.
6. Smith, M., & Morra, J., (2006). *The Prosthetic Impulse: From a Posthuman Present to a Biocultural Future.* Cambridge, MA: MIT Press.
7. Lee, J., (2016). Cochlear implantation, enhancements, transhumanism, and posthumanism: Some human questions. *Science and Engineering Ethics, 22*(1), 67–92. https://doi.org/10.1007/s11948-015-9640-6.
8. Hochberg, L. R., Serruya, M. D., Friehs, G. M., Mukand, J. A., Saleh, M., Caplan, A. H., & Donoghue, J. P., (2006). *Neuronal Ensemble Control of Prosthetic Devices by a Human with Tetraplegia* (Vol. 442, pp. 164–171). Nature Publishing Group. https://doi.org/10.1038/nature04970.
9. Rabins, P., Appleby, B. S., Brandt, J., DeLong, M. R., Dunn, L. B., Gabriëls, L., & Mari, Z., (2009). Scientific and ethical issues related to deep brain stimulation for disorders of mood, behavior, and thought. *Arch Gen Psychiatry, 66*(9), 931–937.
10. Duncan, G. R., (2006). *Catching Seizures Before They Occur – MIT Technology Review.* Retrieved from: https://www.technologyreview.com/s/406069/catching-seizures-before-they-occur/ (accessed on 20 November 2023).
11. Graetz, G., & Michaels, G., (2015). *Robots at Work (No. 1335).* CEP Discussion Paper. London: Center for Economic Performance. Retrieved from: https://ssrn.com/abstract=2575781 (accessed on 20 November 2023).
12. Park, E., & Pobil, A. P. D., (2013). Users' attitudes toward service robots in South Korea. *Industrial Robot: An International Journal, 40*(1), 77–87. https://doi.org/10.1108/01439911311294273.
13. Gui, C., & Chan, V., (2017). Machine learning in medicine. *Medicine & Technology, 86*(2), 76–78. https://doi.org/10.5206/uwomj.v86i2.2060.
14. Lin, L., (1993). *Reinforcement Learning for Robots Using Neural Networks*. Doctoral Dissertation. Carnegie Mellon University.
15. Xu, B., Min, H., & Xiao, F., (2014). A brief overview of evolutionary developmental robotics. *Industrial Robot, 41*(6), 527–533. https://doi.org/10.1108/IR-04-2014-0324.
16. Arvey, I., Paolo, E. Di., Wood, R., Quinn, M., & Tuci, E., (2005). Evolutionary robotics: A new scientific tool for studying cognition. *Artificial Life, 11*(1, 2), 79–98. https://doi.org/10.1162/1064546053278991.
17. Roldan, J. L., & Sanchez-Franco, M. J., (2012). Variance-based structural equation modeling: Guidelines for using partial least squares in information systems research. In: *Research Methodologies, Innovations, and Philosophies in Software Systems Engineering and Information Systems* (pp. 193–221). https://doi.org/10.4018/978-1-4666-0179-6.ch010.
18. Robotics Online, (2018). *Applying Artificial Intelligence and Machine Learning in Robotics.* Retrieved from: https://www.robotics.org/blog-article.cfm/Applying-Artificial-Intelligence-andMachine-Learning-in-Robotics/103 (accessed on 20 November 2023).
19. Bushak, L., (2020). *Robot Surgeon Succeeds Without Help from Human Doctors.* Retrieved from: https://www.newsweek.com/2016/05/20/robot-soft-tissue-surgery-pig-bowels-455765.html (accessed on 20 November 2023).

20. Homburg, N., & Merkle, M., (2019). A cross-country comparison of attitudes toward humanoid robots in Germany, the US, and India. In: *Proceedings of the 52nd Hawaii International Conference on System Sciences* (pp. 4773–4782). Grand Wailea, Maui. https://doi.org/10.24251/HICSS.2019.575.
21. Greguric, I., (2014). Ethical issues of human enhancement technologies: Cyborg technology as the extension of human biology. *Journal of Information, Communication and Ethics in Society, 12*(2), 133–148.
22. Hägele, M., Nilsson, K., Pires, J. N., & Bischoff, R., (2016). Industrial robotics. *Springer Handbook of Robotics*, 1385–1422.

CHAPTER 15

BIOMEDICAL AND HEALTHCARE APPLICATIONS OF NANOROBOTS

P. SURESH[1], U. SARAVANAKUMAR[2], S. SENTHILKUMAR[3], M. SUGANTHI[4], T. ARAVIND[5], and V. K. BUVANESVARI[6]

[1]*Department of Mechanical Engineering, Muthayammal Engineering College, Rasipuram, Namakkal, Tamil Nadu, India*

[2]*Department of ECE, Muthayammal Engineering College, Rasipuram, Namakkal, Tamil Nadu, India*

[3]*Department of EIE, VNE Vignana Jyothi Institute of Engineering and Technology, Hyderabad, Telangana, India*

[4]*Department of ECE, Mahendra College of Engineering, Salem, Tamil Nadu, India*

[5]*Department of CSE, Muthayammal Engineering College, Rasipuram, Namakkal, Tamil Nadu, India*

[6]*Department of MCA, Muthayammal Engineering College, Rasipuram, Namakkal, Tamil Nadu, India*

ABSTRACT

Smaller questions are being asked more frequently by researchers to address some of the most important problems facing the medical field as a whole. Even though most organic techniques operate at the nanoscale, researchers and medical specialists have only recently been able to explore the prospects of nanomedicine thanks to new technological advancements. Featuring us

Robotics and Automation in Healthcare: Advanced Applications. R. Thanigaivelan, Sanjay Singh, & Clement Christy Deepak C. (Eds.)

with top-notch medical degrees, today's infographic showcases some of the most fascinating nanomedicine research.

15.1 INTRODUCTION

Currently, the elderly population is growing worldwide with an increase in comorbidities due to their unhealthy modern habits such as smoking, increased alcohol consumption, unhealthy diet, obesity, and lack of physical activity. A World of Health Organization (WHO) predicts that the health Internet of Things (IoT) industry will explode, costing billions of dollars. IoT startups are gradually introducing modern healthcare applications to integrate network sensors to improve diagnostics, patient monitoring and treatment. Such applications are medical wearables active monitoring of patient data or hospital networks and procedures, optimization providing healthcare and monitoring patient compliance.

IoT is a promising option in healthcare, connecting applications, devices, and people to improve disease treatment with better outcomes and fewer errors, thus ensuring high efficiency and safe healthcare at lower costs Integrating rapidly evolving technologies such as robotics, machine learning (ML), and IoT could change the way of life. Robotics involves machines programmed for labor-intensive work, while ML means that computers and other devices work with no prior programming. Integrating robotics and ML creates robots that work independently. ML techniques range from supervised and semi-supervised learning to unsupervised learning, including statistical pattern recognition, parametric or non-parametric algorithms, neural networks, and other systems. ML in conjunction with IoT could connect many robots, facilitating networking between individuals and things, sending data without intervention, and therefore collecting and monitoring data within networks, for example environmental monitoring, infrastructure management, energy, and production and digitization of buildings, houses, transport, and healthcare. This chapter discusses the applications of adoption of robotics and the Internet of Things in healthcare. It examines the integration of robotics with the Internet of Things for healthcare applications, opportunities, and limitations along with future insights.

Robotics and automation are being utilized more and more in research labs to automate manual, repetitive, and high-volume jobs so that scientists and engineers may concentrate on more strategic work that produces discoveries more quickly. Medical robots have various benefits, including streamlined workflows and risk reduction. Robots, for instance, can independently set up and clean patient rooms, minimizing human touch in infectious disease wards. The time it takes to identify, match, and distribute medications to patients in hospitals is reduced by robots with AI-powered drug recognition software. Robots will become more autonomous as technology develops and eventually be able to complete some activities totally on their own. This will free up more time for physicians, nurses, and other healthcare professionals to directly care for patients.

15.2 BENEFITS OF ROBOTS IN HEALTHCARE

High-quality patient care, effective clinical procedures, and a secure working environment for patients and healthcare professionals are all made possible by the use of robots in the medical area.

15.2.1 EXCELLENT PATIENT CARE

Medical robots can help in the treatment of chronic diseases and provide effective and efficient monitoring of elderly patients. They can also help in the interactions between patients and their caregivers. In addition, they can help improve the quality of life for those with long-term conditions by allowing nurses and other healthcare providers to more effectively interact with patients.

15.2.2 SIMPLIFIED CLINICAL PROCESSES

Independent cell robots (AMRs) make routine tasks simpler, lessen the physical demands on human workers, and ensure more consistent strategies. By keeping track of inventories and placing well-timed orders to help ensure that resources, equipment, and remedies are in stock where they are needed, those robots can solve staffing shortages and demanding situations. Cleaning and sanitizing AMRs make it possible for medical facilities to quickly sanitize and prepare their rooms for new patients, freeing up staff to focus on patient-centered, cost-driven tasks.

15.3 WORKPLACE ENVIRONMENT

AMRs are used to move supplies and linens in hospitals where there is a possibility of disease exposure to keep medical staff safe. Robotic cleaning and disinfection systems help to reduce the spread of pathogens and hospital-acquired infections (HAIs), and hundreds of healthcare facilities currently use them. AMR-style social robots assist with heavy lifting, such as moving patients or beds, relieving physical strain on healthcare professionals.

15.4 SURGICAL ROBOTS

Surgical-assistance robots have become increasingly distinctive as movement manipulation technology has developed. These surgical robots assist surgeons in achieving unprecedented levels of speed and precision while carrying out difficult surgeries using AI- and computer vision-successful technology. Some surgical robots may even be able to do tasks on their own, enabling physicians to oversee procedures from a console (Figure 15.1).

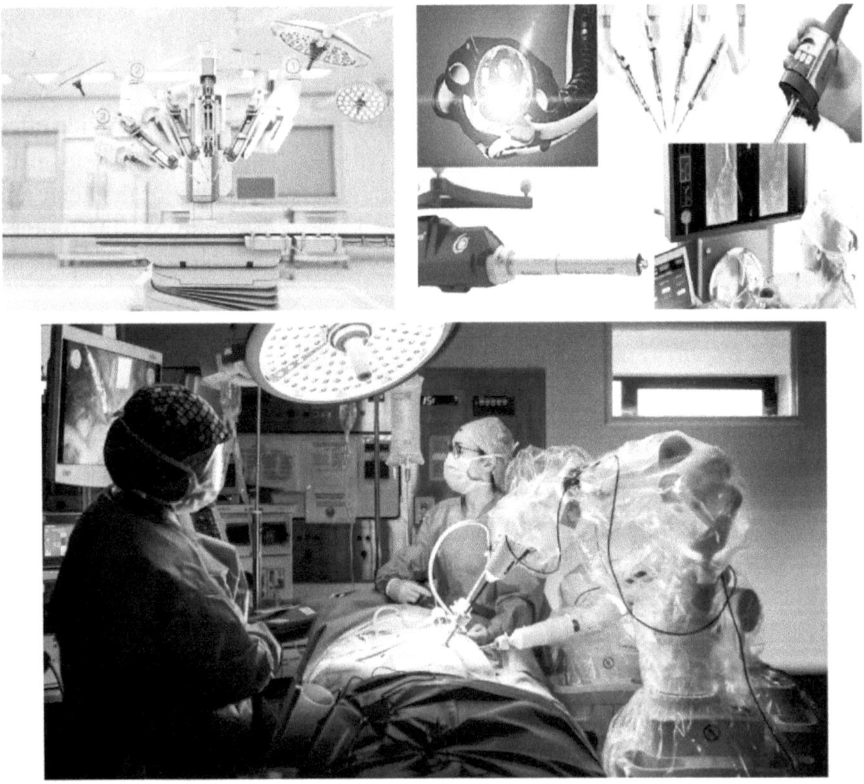

FIGURE 15.1 Surgical-assistance robots.

Source: Images compiled by authors

Surgeries performed with robotics assistance fall into two main categories:

1. **Minimally Invasive Torso Procedures:** Robotic hysterectomy, robotic prostatectomy, bariatric surgery, and other techniques primarily aimed at smooth tissues are some of them. These robots are inserted through a small incision and then lock into place, creating a sturdy platform from which to perform surgical procedures using remote control. Open surgical procedures involving large incisions were formerly common for the majority of internal functions. Recovery times have increased significantly, and contamination and other issues are now more likely. Even for a skilled medical professional, physically operating via a button-sized incision is quite difficult. In order to reduce infections and other complications, surgical robots make these procedures precise and streamlined (Figure 15.2).

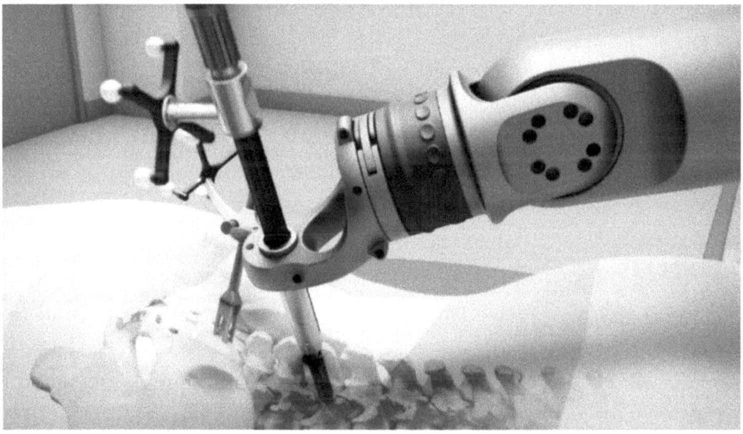

FIGURE 15.2 Orthopedic surgery robot.

Source: Still from video by Globus Medical on YouTube. https://www.youtube.com/watch?v=ah-xUUIrGNU&t=13s

2. **Orthopedic Surgeries:** Devices may be pre-programmed to do common orthopedic procedures like hip and knee replacements. To assist the physician, these robots use spatially specified obstructions and integrate robotic hands with three-dimensional (3D) imaging and data analytics to produce more predictable results. AI modeling can be used to teach robots precise orthopedic surgery skills, including the optimum locations to go and how to complete each step. Surgeons can learn from consultations with other experts in their area because of the ability to broadcast a video clip from the operating room to

sites both close by and far away. Patients are worried about the doctors' methods as a result.

The use of artificial intelligence (AI) in surgical robotics is growing. Computer vision enables surgical robots to discern between various tissue types in their range of view. For example, surgical robots now help doctors avoid nerve and muscle tissues during various surgeries. High-quality 3D during operations, computer vision can give surgeons specialized information and improve performance. In the future, robots might be able to perform minor procedures like suturing or other tasks under the close supervision of a medical professional. In addition, robotics is important for training medical professionals. AI and virtual reality are combined in simulation platforms to teach surgical robots. Surgeons can practice techniques and hone abilities using robotic controls in a digital environment.

15.5 MODULAR ROBOTS

Modular robots can be set up to perform a few different tasks and improve other systems. These include prosthetic robotic legs and arms and healing exoskeleton robots in the field of healthcare. Robotic therapy can aid in the recovery of patients suffering from multiple sclerosis-related difficulties, severe brain injuries, paralysis, or strokes. Intel and Accenture are presently working on a robotic arm that can be installed on a wheelchair to help persons with spinal injuries perform daily chores. While a patient performs the exercises suggested, robots with AI and intensity cameras may monitor their shape, detecting movements in certain positions and monitoring progress more precisely than the human eye can. They can interact with patients to encourage them and provide knowledge.

15.6 AUTONOMOUS MOBILE ROBOTS

While AMRs can assist with essential tasks like disinfection, telepresence, and transportation of medication and medical supplies, they are frequently utilized by healthcare organizations to help create secure settings and free up staff members to spend more time with patients. When equipped with light detection and varying (LiDAR) structures, visible computation, or mapping capabilities, AMRs can self-navigate to patients in exam or medical center rooms, enabling doctors to converse from a distance. An AMR that is managed by a remote specialist or another member of staff can accompany a physician when they are performing rounds at a hospital and allow the specialist to take part in a discussion about the diagnosis and treatment of patients via an online session.

Just before patient is checked in, some robots can assist medical workers. For instance, a self-sustaining robot created with assistance from the startup Roomie is used in Mexico to support medical workers with high-chance patients. COVID-19 sufferers. A robot named RoomieBot triages patients as they arrive at the hospital by taking their temperature, blood oxygen level, and clinical data. AI algorithms and other Intel-based technology are used by RoomieBot (Figure 15.3).

15.7 MEDICAL ROBOTS

15.7.1 ROBOT-ASSISTED SURGERY

Robots are frequently used in the medical industry to support healthcare workers in their regular tasks. Automating repetitive chores in a research lab, navigating hospitals on its own to carry linens to patient rooms, and keeping an eye on a patient's form as they perform physical therapy exercises are all examples of how robots might aid.

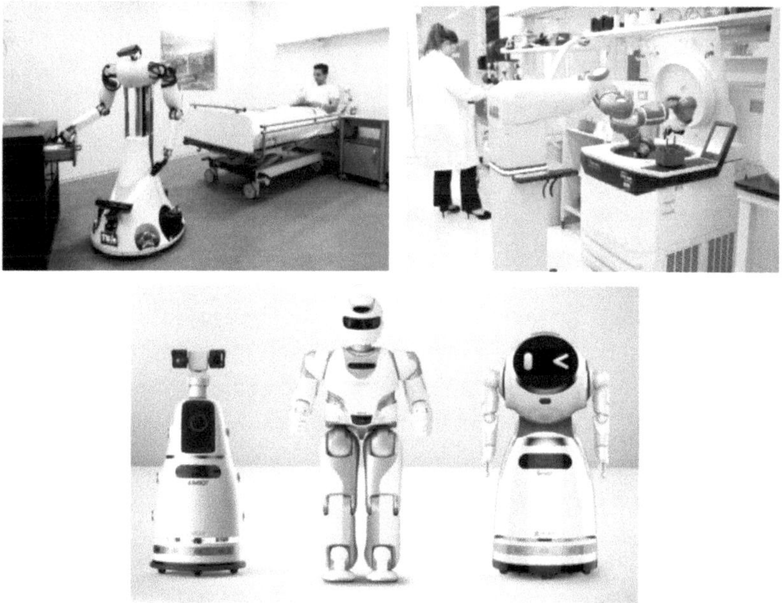

FIGURE 15.3 Autonomous mobile robots.

Source: Compiled by authors

Since the 1980s, robots have been used in medicine, initially to help doctors in the operating theater. Robots performing surgery are becoming a more common occurrence in operating rooms today. The use of robots in addition to surgery has some clear benefits. Surgeons and surgical teams receive dependable, continuous intraoperative assistance from robots. For instance, in spine surgery, robots can keep instruments and implant parts perfectly still and in place while placing screws for decompression surgery. Beyond routine methods, robotics can also help by aiding the crew during difficult tactics, such as single-function fusions, which frequently call for unergonomic setups for the healthcare professional. Robotics, when used in conjunction with surgical navigation during scoliosis surgery, can help with the alignment issues these cases' complex anatomical structures bring. This secure positioning of the devices not only enables higher precision, but also frees up the hands of the medical and surgical teams at various points in the procedure. Robots can reduce the overall technique time of surgical procedures because of their dependability. A general surgeon must drill while continuously verifying and maintaining the right alignment when performing backbone surgery without the use of a robot. With the use of a surgical robotic, alignment and drilling are divided into two successive processes, reducing the cognitive load on the physician, and perhaps enabling them to finish drilling more quickly. The hospital, the surgeon, and especially the patient benefit from quicker operations because they may require less time for anesthesia.

15.7.2 ROBOTICS FOR RADIOTHERAPY

Robotics were first used in the radiation industry in the 1990s. The main device had a linear accelerator attached to a robot arm that could move about the body and precisely treat tumors in various places. Since then, robotics have been applied in radiosurgery and radiotherapy in a similar manner. Robotic therapy couches, for instance, precisely position the patient before the commencement of the treatment. They also enable clinicians to move the patient from a distance without entering the treatment area.

15.7.3 REHABILITATION ROBOTS

Although the area of rehabilitation robots is relatively young, Theodor Büdingen developed the idea of utilizing machines to help patients recover as early as 1910 when he invented a device to support walking movements in heart disease patients (Figure 15.4).

The first genuine robot devices created for rehabilitation were based on the principles of non-stop passive movement (CPM), which moves a portion of the patient's body while they relax. Gait rehabilitation is an example of this use. Robotics offer controlled, repetitive, and intense training compared to traditional physical therapy methods, which can lighten the load on the therapist and enable quantitative evaluation of the patient's improvement. Due to these advantages, robots are being used more and more frequently in rehabilitation.

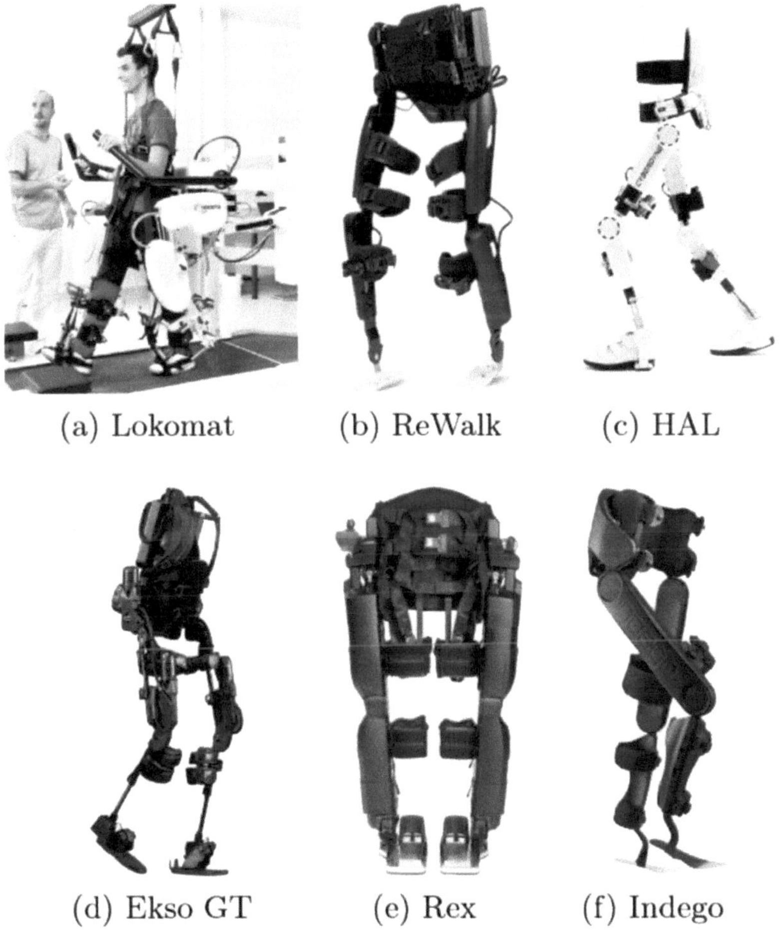

FIGURE 15.4 Rehabilitation robots.

Source: Reprinted with permission from Ref. [52]. Copyright 2020. Springer Nature

15.7.4 LABORATORY ROBOTS

Robots have been a common sight in labs for the past 30 years. The kinds of robots that have been found in laboratories are particularly made to automate processes or aid technicians in completing monotonous duties (Figure 15.5).

FIGURE 15.5 Laboratory robot.

Photo courtesy: ABB Robotics, reprinted from Assembly. https://www.assemblymag.com/articles/95814-cobot-ensures-faster-sample-analysis-in-the-lab

Like in the field of commercial robotics, laboratory robots frequently do tasks involving chemicals and substances that are hazardous or dangerous to humans. The automation laboratory robots enable will improve in speed, aptitude, and accuracy by reducing human errors.

15.7.5 ROBOTIC PROSTHETICS

The shockingly modern software for medical robots specializes in giving wearers the capacity to move their limbs like real people. Although there are now robotic prostheses on the market, as this period develops, they continue to be quite expensive for patients. Neuromusculoskeletal prosthesis, which can be attached to the bone and are controlled by bidirectional interfaces connected to the patient's neuromuscular machine with electrodes inserted in their nerve and muscle tissue, are one example of breakthroughs in this field. A prosthesis, often known as a prosthetic limb, is a device that acts as a missing limb replacement. Although there are many other types of prosthetics available, a robot prosthetic is by far the most advanced. Robotic prosthetics are artificial limbs that a human control with microprocessors and nerve impulses, a far cry from the timber or steel fingers of bygone times. Here is a brief summary of what you might discover while looking for prosthetics in Memphis (Figure 15.6).

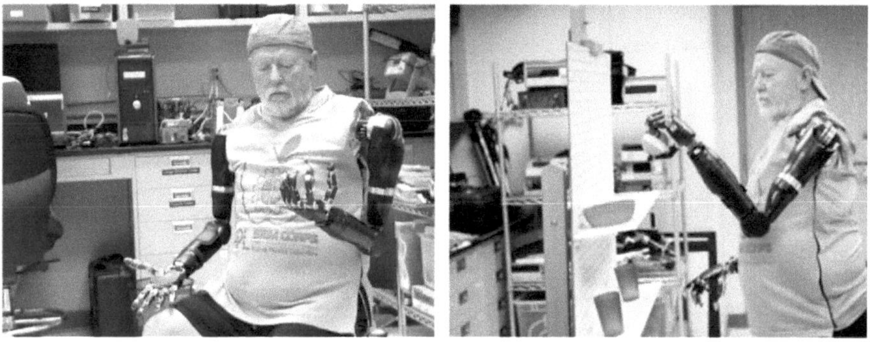

FIGURE 15.6 Robotic prosthetics arm.
Source: Stills from video by Johns Hopkins University Applied Physics Laboratory

15.7.6 TYPES OF LIMBS

While doing your initial study on this subject, you could feel overwhelmed by the choices offered by several businesses, each of which is attempting to make their product as user-friendly and pleasant as possible. For instancc, the iWalk firm has a foot that constantly uses algorithms to generate different pressure levels, turn the foot at the proper angles, and so forth depending on the terrain. A knee from Ossur is available to analyze your foot sample in order to be more effective. Some organizations currently possess robotic arms with digits that can be powered independently for stepped forward motor control. (Recent models use springs and a cable-conduit device rather than electric automobiles, making them lighter than vintage ones.) Before making a decision, consider all of your options because there are numerous factors to consider while selecting a new limb.

15.7.7 SENDING SIGNALS

Finding a mechanism for the user to instruct the robotic limb to perform their preferred movements is one of the most challenging aspects of creating a robotic limb. Scientists connect the nervous system to the electronic components of the prosthetic in a variety of unique methods. Electromyography sensors are widely used for this, which periodically check in as the user contracts their muscles and moves their arm. As an alternative, in targeted muscle reinvention, a doctor performs surgery on the patient and rewires the nerves in any other area of the body to run the robot limb. Surgically implanting sensors into the severed stump is a third technique to transmit indicators. These sign-sending methods can all be challenging. Most prosthetic limb users must concentrate on their movements, and occasionally they are unable to perform any other tasks while transferring that limb.

15.7.8 THE FUTURE OF MEDICAL ROBOTICS

In every case, the "distance to the patient" tends to increase with the amount of autonomy of healthcare robots. For instance, a surgical robot that is placed very close to the patient but lacks autonomy is directed exactly how to behave by the doctor. However, sanitation robots, which may be completely phased out from working with patients in the near future, have a greater capacity to decide how to behave based solely on their environment. There are several ways that robots are used in healthcare today, many of which are intended to help people in ways that go beyond what we can clearly and effectively do on our own. The programs of those types of robots continue to extend inadvertently in the realms of medicine and surgery. Robots in operating rooms and clinics are already becoming commonplace and are just one of many ways that healthcare continues to push the boundaries of technology.

15.8 NANOTECHNOLOGY IN BIOMEDICAL

Certainly, nanotechnology has existed for a long. Richard Feynman, a Nobel Prize winner, made a significant contribution to the field when he told a group of physicists at the American Physical Society meeting at Caltech in 1959 that "there may be plenty of room down the bottom." In his essay, Feynman—often referred to as the father of nanotechnology—described a theoretical approach that would let scientists work with single atoms or single molecules. Although it wasn't conceived, this system may ultimately become the main application of nanoscience. Robotics may be summed up as the idea and attentiveness of robots, which are relatively autonomous digital, electric-powered, or mechanical devices that are used for tasks like manufacturing. You can speak about nanotechnology robotics, or, to put it briefly, nanorobotics, if you scale that robotic down to 3 billionths of a meter. The field of nanorobotics unites a number of specialties, including nanofabrication techniques used to create nanomotors, nanoactuators, nanosensors, and physical modeling at nanoscales.

A component of nanorobotics is nanorobotic manipulation technology, which includes the assembly of nanometer-sized parts, the manipulation of organic cells or molecules, and the types of robots used to carry out these tasks. For instance, researchers have modified traditional static optical fabrication systems by translating the self-sufficient motion trajectories of nanomotors into controlled floor capabilities. This offers an early degree technique for a nanorobotics platform for nanomanufacturing.

Nanotechnology is more than just science fiction villains' technology; it's a rapidly developing field that combines engineering and science. The term "nanorobotics" refers to the developing field of designing and creating robots whose parts are composed of nanoscale or molecular additives and are roughly the size of a nanometer (109 meters) or range in size from 0.1 to 10 micrometers. Nanotechnology robots are classic examples of nano-electromechanical systems (NEMS), and they address all the crucial issues that must be taken into consideration when designing NEMS, including sensing, actuation, control, communications, energy, and interfacing across spatial scales and between natural and inorganic materials. Nanorobots have a wide range of capability programs in areas like environmental monitoring or medical due to their length, which is equivalent to that of biological cells. However, today's researchers typically use the term "nanorobot" to describe molecules that have special properties that allow them to be trained to carry out a specific task. Nanoadjuvants having immunomodulatory qualities that are used to deliver vaccine antigens, the nano-knife, an almost non-invasive means of eliminating most malignancies cells with high voltage strength, and carbon nanotubes are some more possible applications of nanotechnology in medicine. Packages of nanoparticles that are now being developed as well as longer-range research that uses manufactured nanorobots to do maintenance on a mobile level are both examples of applications of nanotechnology in medicine. The use of nanotechnology in the medical profession should change how we diagnose and treat diseases and injuries to the human body in the future, and some previously unimagined methods are already making amazing strides toward becoming a reality. For comparison, one nanometer is about equivalent to 10 times the size of a single atom and 10 times smaller than the width of your DNA (Figure 15.7).

The development of microscopes that could even view individual atoms didn't happen until 1981. Early scanning tunneling microscopes achieved a level of accuracy and magnification not previously seen. By enabling scientists to visualize individual atoms, they contributed to the idea that nanotechnology became feasible (Figure 15.8).

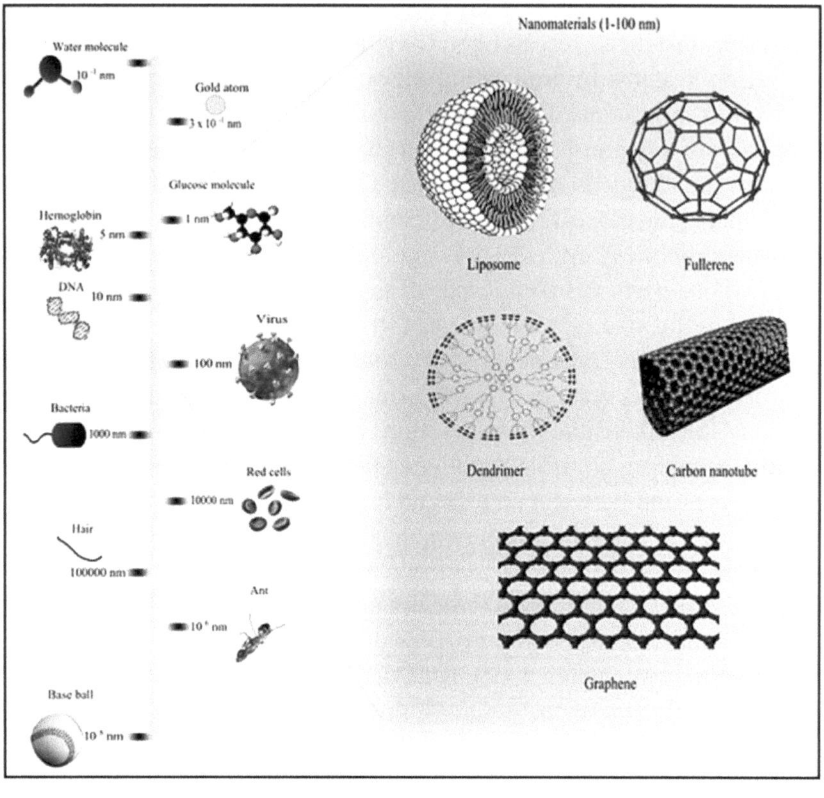

FIGURE 15.7 A scale of nanotechnology matched up with other understandable objects.

Source: Reprinted with permission from Ref. [51]. Copyright © 2014 by the authors; licensee MDPI, Basel, Switzerland. http://creativecommons.org/licenses/by/3.0/

15.9 NANOTECHNOLOGY IN MEDICINE APPLICATION

15.9.1 DRUG DELIVERY

One currently under development use of nanotechnology in medicine is the delivery of medications, heat, light, or other chemicals to particular types of cells via nanoparticles (including most cancers cells). Debris is made to draw ill cells, enabling direct therapy of specific cells. Early illness identification is possible with this technique, and healthy cells in the body sustain less damage.

For instance, researchers at the Kingdom College of North Carolina are developing a method to transplant cardiac stem cells to heart tissue that has been injured. In order to transfer more stem cells to an injured area, they link nanovesicles that are concerned with stem cell injury. Massive length fabrics used for drug administration have problems such as poor bioavailability, low solubility, ineffective transport, and wide-ranging adverse effects. The use of nanotechnology in medicine delivery makes it possible to deliver medications with greater precision and with fewer negative effects. Nanotechnology medication delivery programs are developed by the use of specifically designed nanomaterials and the creation of transport structures from nanoscale molecules like liposomes.

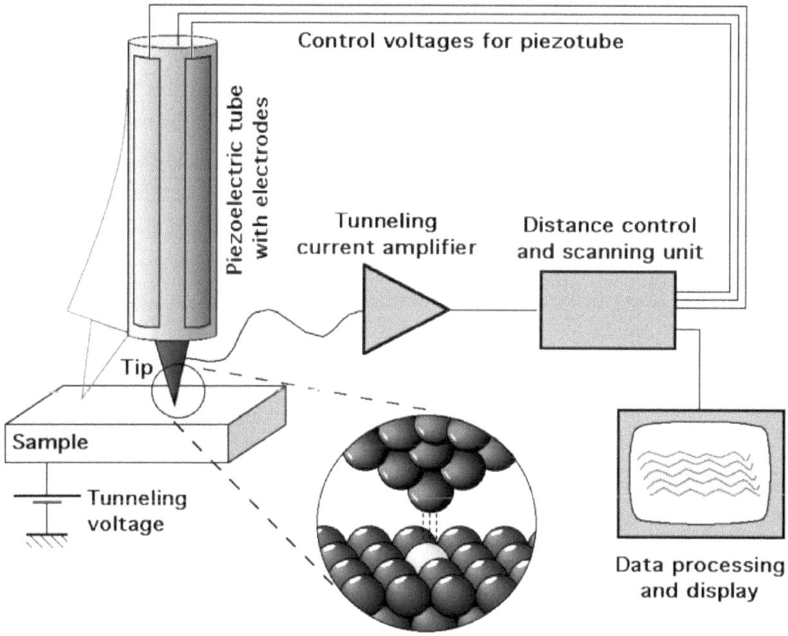

FIGURE 15.8 A schematic diagram of how a scanning tunneling microscope works.

Image courtesy: Michael Schmid and Grzegorz Pietrzak. https://creativecommons.org/licenses/by-sa/2.0/at/deed.en

Using nanotechnology for drug delivery must result in the following advantages:

- Increase the ability to provide potentially water-insoluble medications.

- Provide site-specific targeting to reduce medication buildup inside healthy tissue.
- Assist in keeping the medication in the body long enough for effective treatment.
- The expansion of medication bioactivity via organic environment safety.
- Permit the passage of medicine via endothelial and epithelial barriers.
- Consolidate therapeutic and diagnostic approaches into a single agent.

15.9.2 THE ENHANCED PERMEABILITY AND RETENTION EFFECT (EPR)

The enhanced permeability and retention impact is a crucial finding for medication transportation nanotechnology efforts. Nanoscale molecules tend to accumulate in tumor tissue more so than in normal tissue. This is due to the rapid development of blood vessels and the high-quality oxygen demand of rapidly forming tumors. Strange structural features and pores in the newly created blood arteries allow nanoscale chemicals to pass through to the tumor tissue. The increased permeability and retention effect must enable targeted delivery of chemotherapy tablets because rapidly growing cancers also lack functional lymphatic networks.

Through the improved permeability and retention effect, studies have looked at which types of cancers have reacted very well to selective drug release. According to statistics from scientific trials, the polymer drug conjugate evaluated has the best first-class response with increased expression in lung, breast, and ovarian malignancies. The findings showed that those tumor types had significantly greater retention and permeability effects. It has been suggested that selecting patients based on their higher permeability and retention impact will also increase the medical capabilities of this drug delivery system.

15.9.3 NANOTECHNOLOGY DRUG DELIVERY FOR INFECTION TREATMENT

Nanotechnology is being used to combat contamination caused by residues of germs that are becoming drug-resistant while drugs are being sent. Although the increased permeability and retention effect is mostly used in cancer treatments, the impact can also be used in pollution treatments due to overlapping pathophysiological pathways. Vasodilation quickly occurs at some point during contamination, increasing the permeability of capillaries.

Even though the retention effect in cancer tumors is due to the absence of a functioning lymphatic system, studies have shown that after an infection, fast lymphatic clearance of small molecules does not always occur as anticipated. In reality, ineffective lymphatic drainage is a symptom of infection. Nanoantibiotics will take advantage of this effect to enhance infection therapy through the employment of both nanoparticles with antimicrobial healing mechanisms and nanosized biological molecules for advanced antibiotic medicine delivery.

15.9.4 NANOTECHNOLOGY DRUG DELIVERY FOR TREATMENT OF HEART DISEASE

The development of drug delivery systems for heart disease is another area where nanotechnology is being applied. Because of the way that inflammatory changes also result in increased vascular permeability and retention of nanoscale molecules, targeted medication delivery to sick coronary heart tissue is possible. Coronary artery disease is the cutting-edge plaque buildup at the significant arteries that can result in coronary heart failure. Because the illness develops at the cellular level, nanomedicine can be used as a valid treatment. Modern treatment for coronary artery disease mostly involves the use of statins, which are known to have dose-dependent side effects that restrict their use. Nanotechnology could enable the targeted delivery of statins to the site of need, reducing their toxicity to various cell types:

1. **Diagnostic Techniques:** In order to develop a sensor that could quickly identify COVID-19 and other viruses, researchers at John Hopkins University are employing nanoimprint lithography. Researchers at Worcester Polytechnic Institute are employing antibodies connected to carbon nanotubes embedded in chips to detect the majority of cancer cells in blood flow. The approach may be employed in simple lab tests to detect cancer cells in the circulation early, according to the researchers.

   The development of a test for the early detection of kidney disease has begun. Gold nanorods that have been functionalized to connect to the specific protein made by damaged kidneys are used in the approach. The nanorod's color changes as protein deposits on it.

The test is designed to be simple to use and inexpensive for problem early identification.

2. **Antibacterial Treatments:** Infrared light and gold nanoparticles are being developed by researchers at the University of Houston to kill microorganisms. This approach might also result in more thorough cleaning of equipment in medical facilities. At the University of Colorado Boulder, researchers are looking into the use of quantum dots to treat illnesses that are resistant to antibiotics.
3. **Wound Treatment:** University of Wisconsin researchers have authorized a bandage that delivers electric pulses to a wound using electricity produced by the wearer's own nanogenerators. In order to stop the blood loss in trauma victims who have internal bleeding, all other available options must be used. Researchers from Case Western Reserve College are working to create polymer nanoparticles that can simulate platelets. Injecting these artificial platelets considerably reduces blood loss, according to laboratory testing.

15.10 THE FUTURE OF NANOTECHNOLOGY IN MEDICINE

Researchers are increasingly asking smaller questions to address some of the most significant issues in medicine all around the industry. Although the majority of organic strategies take place at the nanoscale, it wasn't until recently that new technology developments helped open up the possibilities of nanomedicine to researchers and specialists in the field of medicine. Today's infographic, which includes us with high-caliber health degrees, displays some of the most exciting nanomedicine studies.

KEYWORDS

- **artificial intelligence**
- **hospital-acquired infections**
- **internet of things**
- **machine learning**
- **three-dimensional**
- **World of Health Organization**

REFERENCES

1. Senthilkumar, S., & Suresh, P., (2015). Analysis of virtual robot movements for performance improvements in personal communication. *Wireless Personal Communication, 90*(2), 595–608. ISSN: 0929-6212 (Print) 1572-834X (Online). Springer.
2. Senthilkumar, S., & Suresh, P., (2015). Improving the leg movement in locomotion hexapod robot for industrial automation. *International Journal of Advanced Engineering Research, 10*(20), 41377–41387. ISSN 0973-4562 (Print), ISSN 1087–1090 (Online).
3. WHO, (2020). *Report of the WHO-China Joint Mission on Coronavirus Disease 2019 (COVID-19).* Geneva, Switzerland: WHO.
4. Fanelli, D., & Piazza, F., (2020). *Analysis and Forecast of COVID-19 Spreading in China, Italy, and France* (Vol. 134, p. 109761). Chaos, Solitons & Fractals. doi: 10.1016/j.chaos.2020.109761.
5. Gostic, K., Gomez, A. C., Mummah, R. O., Kucharski, A. J., & Lloyd-Smith, J. O., (2020). Estimated effectiveness of symptom and risk screening to prevent the spread of COVID-19. *eLife*, 9, article e55570. doi: 10.7554/eLife.55570.
6. Kannan, S., Ali, P. S. S., Sheeza, A., & Hemalatha, K., (2020). COVID-19 (novel coronavirus 2019)-recent trends. *European Review for Medical and Pharmacological Sciences*, *24*(4), 2006–2011. doi: 10.26355/eurrev_202002_20378.
7. Mijares, S. F., & Chan, P., (2018). Ethical robots in healthcare? *Journal of Academy of Business and Economics™*, *18*(3), 5–16. doi: 10.18374/JABE-18-3.1.
8. Alotaibi, M., & Yamin, M., (2019). 2019 6th International Conference on Computing for Sustainable Global Development (INDIACom) IEEE. *Role of Robots in Healthcare Management* (pp. 1311–1314).
9. Kim, J., Gu, G. M., & Heo, P., (2016). *Biomedical Engineering: Frontier Research and Converging Technologies* (pp. 489–509). Cham: Springer; Robotics for Healthcare.
10. Coltart, C. E., Lindsey, B., Ghinai, I., Johnsonand, A. M., & Heymann, D. L., (2017). The Ebola outbreak, 2013–2016: Old lessons for new epidemics. *Philosophical Transactions of the Royal Society, B: Biological Sciences*, *372*(1721), 20160297. doi: 10.1098/rstb.2016.0297.
11. Tavakoli, M., Carriere, J., & Torabi, A., (2020). Robotics, smart wearable technologies, and autonomous intelligent systems for healthcare during the COVID-19 pandemic: An analysis of the state of the art and future vision. *Advanced Intelligent Systems.*
12. Romero, M., Huerfano, L. M., & Melo, E. V., (2019). PNS16 evaluation of the multicriteria methodology for the use in evaluation of health technologies. Advantages and disadvantages of the method. *Value in Health, 22*, S290. doi: 10.1016/j.jval.2019.04.1382.
13. Patel, A. R., Patel, R. S., Singh, N. M., & Kazi, F. S., (2017). Internet of things and big data technologies for next-generation healthcare. *Vitality of Robotics in Healthcare Industry: An Internet of Things (IoT) Perspective* (pp. 91–109). Cham: Springer.
14. Kanzawa, M., Spindler, H., Anglemyer, A., & Rutherford, G. W., (2020). Will coronavirus disease 2019 become seasonal? *The Journal of Infectious Diseases*, *222*(5), 719–721. doi: 10.1093/infdis/jiaa345.
15. Chen, B., Marvin, S., & While, A., (2020). Containing COVID-19 in China: AI and the robotic restructuring of future cities. *Dialogues in Human Geography*, *10*(2), 238–241. doi: 10.1177/2043820620934267.
16. Yoganandhan, A., Kanna, G. R., Subhash, G. S. D., & Jothi, J. H., (2021). Aplicacion retrospectiva y prospectiva de robots e inteligencia artificial en pandemias y epidemias globales. *Vacunas (English edn.)*, *22*(2), 98–105. doi: 10.1016/j.vacune.2020.12.002.

17. Khan, Z. H., Siddique, A., & Lee, C. W., (2020). Robotics utilization for healthcare digitization in global COVID-19 management. *International Journal of Environmental Research and Public Health, 17*(11), 3819. doi: 10.3390/ijerph17113819.
18. Sahasranamam, S., (2020). *How Coronavirus Sparked a Wave of Innovation in India.* World Economic Forum.
19. Seeja, G., Reddy, O., Korupalli, V., Kumar, R., & Mounika, S. S. L. C. H., (2021). Innovations in the industrial internet of things (IIoT) and smart factory. *Internet of Things and Robotic Applications in the Industrial Automation Process* (pp. 50–64). IGI Global.
20. Farkh, R., Marouani, H., Al Jaloud, K., Alhuwaimel, S., Quasim, M. T., & Fouad, Y., (2021). Intelligent autonomous robot control for medical applications. *Computers, Materials, and Continua, 68*(2), 2189–2203. doi: 10.32604/cmc.2021.015906.
21. Bajpai, N., Biberman, J., & Wadhwa, M., (2020). *ICT Initiatives in India to Combat COVID-19.* Columbia Academic Commons.
22. Aymerich-Franch, L., & Ferrer, I., (2020). *The Implementation of Social Robots During the COVID-19 Pandemic.* https://arxiv.org/abs/2007.03941.
23. Gharpure, R., Hunter, C. M., Schnall, A. H., et al., (2020). *Knowledge and Practices Regarding Safe Household Cleaning and Disinfection for COVID-19 Prevention—United States.*
24. Ramalingam, B., Yin, J., Rajesh, E. M., et al., (2020). A human support robot for the cleaning and maintenance of door handles using a deep-learning framework. *Sensors, 20*(12), 3543. doi: 10.3390/s20123543.
25. Romero, M. C., (2021). *Development of an AGV Robot Based on ROS for Disinfection in Clinical Environments.*
26. Ackerman, E., (2020). *Autonomous Robots are Helping Kill Coronavirus in Hospitals.* IEEE Spectrum.
27. Puri, O., Rathaur, V. K., Pathania, N., & Pathania, M., (2020). A new phase of healthcare: COVID-19 and medical advancements. *Journal of Clinical and Diagnostic Research, 14*(11). doi: 10.7860/JCDR/2020/46510.14254.
28. Macalam, M., & Locsin, R., (2020). Humanoid nurse robots and compassion: Dialogical conversation with Rozzano Locsin. *Journal of Health and Caring Sciences, 2*(1), 71–77. doi: 10.37719/jhcs.2020.v2i1.rna001.
29. Kim, D. K. D., Kreps, G., & Ahmed, R., (2021). Communicative development and diffusion of humanoid AI robots for the post-pandemic healthcare system. *Human-Machine Communication, 3*(1), 65–82. doi: 10.30658/hmc.3.5.
30. Giorgi, I., Watson, C., Pratt, C., & Masala, G. L., (2021). Human-centered intelligent systems. *Designing Robot Verbal and Nonverbal Interactions in Socially Assistive Domain for Quality Ageing in Place* (pp. 255–265). Singapore: Springer.
31. Almalki, F. A., & Soufiene, B. O., (2021). EPPDA: An efficient and privacy-preserving data aggregation scheme with authentication and authorization for IoT-based healthcare applications. *Wireless Communications and Mobile Computing.*
32. Mukherjee, U. K., & Sinha, K. K., (2020). Robot-assisted surgical care delivery at a hospital: Policies for maximizing clinical outcome benefits and minimizing costs. *Journal of Operations Management, 66*(1, 2), 227–256.
33. Edelman, L. S., McConnell, E. S., Kennerly, S. M., Alderden, J., Horn, S. D., & Yap, T. L., (2020). Mitigating the effects of a pandemic: Facilitating improved nursing home care delivery through technology. *JMIR Aging, 3*(1, article e20110). doi: 10.2196/20110.

34. Kaur, A., Mittal, N., Khosla, P. K., & Mittal, M., (2021). Predictive and preventive measures for COVID-19 pandemic. *Machine Learning Tools to Predict the Impact of Quarantine* (pp. 307–323). Singapore: Springer.
35. Chauhan, A., (2021). Artificial intelligence for COVID-19. *Robotics and Automation: The Rescuers of COVID Era* (pp. 119–151). Cham: Springer.
36. Malik, A. A., Masood, T., & Kousar, R., (2020). Repurposing factories with robotics in the face of COVID-19. *Science Robotics*, *5*(43). doi: 10.1126/scirobotics.abc2782.
37. Mahdy, L. N., Ezzat, K. A., Darwish, A., & Hassanien, A. E., (2021). *The Role of Social Robotics to Combat COVID-19 Pandemic* (pp. 205–217).
38. Kaminski, J., (2020). Informatics in the time of COVID-19. *Can J Nurs Inform.*, *15*(1). [Google Scholar].
39. Podpora, M., Gardecki, A., Beniak, R., Klin, B., Vicario, J. L., & Kawala-Sterniuk, A., (2020). Human interaction smart subsystem—Extending speech-based human-robot interaction systems with an implementation of external smart sensors. *Sensors*, *20*, 2376. doi: 10.3390/s20082376.
40. Sierra, M. S. D., Gomez-Vargas, D., Céspedes, N., et al., (2021). Expectations and perceptions of healthcare professionals for robot deployment in hospital environments during the COVID-19 pandemic. *Frontiers in Robotics and AI*, *8*, 102. doi: 10.3389/frobt.2021.612746.
41. Murphy, R. R., Gandudi, V. B. M., & Adams, J., (2020). *Applications of Robots for COVID-19 Response.* https://arxiv.org/abs/2008.06976.
42. Chen, C., Demir, E., Huang, Y., & Qiu, R., (2021). The adoption of self-driving delivery robots in last-mile logistics. *Transportation Research Part E-Logistics and Transportation Review*, *146*, 102214. doi: 10.1016/j.tre.2020.102214.
43. Plischke, S., Machutova, J., Stasa, P., & Unucka, J., (2020). *2020 IEEE 2nd Eurasia Conference on Biomedical Engineering, Healthcare and Sustainability (ECBIOS)* (pp. 76–80). IEEE. Development of software interface for transfer of patient medication preparation from Czech DASTA standard to international HL7 standard.
44. Wolter, R., Hindriks, K. V., Samur, D., & Jonker, C. M., (2020). International conference on practical applications of agents and multi-agent systems. *A Study on Automated Receptionists in a Real-World Scenario* (pp. 340–352). Cham: Springer.
45. Dao, N., Hai, X., Huu, L., Nam, T., & Thinh, N. T., (2019). International conference on system science and engineering (ICSSE) IEEE. *Remote Healthcare for the Elderly, Patients by Tele-presence Robot* (pp. 506–510).
46. https://www.ft.com/content/2c47aaba-29e3-4f6a-b1ef-6037fa68513d (accessed on 20 November 2023).
47. https://www.ul.com/news/robots-surgical-assistance (accessed on 20 November 2023).
48. https://www.therobotreport.com/7-companies-transforming-orthopedics-with-robot-assisted-surgery/ (accessed on 20 November 2023).
49. https://www.automate.org/blogs/robotics-in-human-rehabilitation (accessed on 20 November 2023).
50. https://en.wikipedia.org/wiki/Scanning_tunneling_microscope (accessed on 20 November 2023).
51. Panneerselvam S, Choi S. Nanoinformatics: Emerging Databases and Available Tools. International Journal of Molecular Sciences. 2014; 15(5):7158-7182. https://doi.org/10.3390/ijms15057158

52. Cardona, M., Destarac, M., Cena, C. (2020). Robotics for Rehabilitation: A State of the Art. In: Exoskeleton Robots for Rehabilitation and Healthcare Devices. SpringerBriefs in Applied Sciences and Technology. Springer, Singapore. https://doi.org/10.1007/978-981-15-4732-4_1

INDEX

A

Abdomen, 138, 243
Accelerometer, 88, 141, 152
Accuracy, 53, 63, 65, 66, 68, 69, 78, 95, 97, 100, 105, 106, 109, 110, 116, 130, 132, 145, 159, 170, 172, 223, 235, 266, 270
Acrylonitrile butadiene styrene (ABS), 28, 77, 95, 98, 100, 107–109, 111
Actuators, 26, 137–146
Adaptability, 220, 228
Aerial vehicles, 73, 187
Aerodynamics, 189, 190, 193, 196
Aerospace industries, 105, 137
Amplify, 123, 244
Analysis, 24, 46, 47, 51, 65, 106, 116, 120, 121, 128, 130, 133, 144, 145, 155, 171, 172, 200, 225, 227, 266
Analytical, 46, 171, 184
  hierarchy process (AHP), 169, 171, 178, 179, 183, 184
Anatomy, 121, 132
Angle of attack (AOA), 189, 197
Antenna, 2, 3, 85, 161, 244
Arduino, 25, 156, 159–161, 163, 166, 208, 212–214, 217, 218, 220
  board, 156, 159, 160, 163, 166, 214
Arousal, 239, 254
Artificial
  intelligence (AI), 22, 23, 25, 43, 129, 130, 132, 246, 259–263, 274
  limbs, 267
  neural network (ANN), 24, 43, 105, 107, 110, 226, 236, 248
Asynchronous actor-critic algorithm (A3C), 248
Atmospheric flight, 193, 194
Augmented reality (AR), 132, 134
Automated external defibrillator (AED), 77, 92
Automation, 25, 31, 62, 77, 137, 145, 146, 156, 166, 169, 207, 210, 246, 258, 266
Autonomous
  flight, 78, 86
  mental development (AMD), 249
  mobile robots, 262
  robot, 21
  robotic surgery, 115
Autotitrating (variable) positive airway pressure (APAP), 209
Average variance extracted (AVE), 251, 252
Axes, 194–196, 246

B

Barium titanate (BaTiO), 127
Barriers, 65, 189, 272
Bilevel positive airway pressure (BiPAP), 209, 221
Biological cells, 140, 269
Biologically galvanized methods, 65
Biomechanical study, 46, 48
Biomechanics, 46, 48
Biosignals, 152, 166
Block diagram, 200, 212, 213
Bone
  block, 45, 49–51, 58, 59
  implant interface, 46, 47, 49
  properties, 48, 50
Bony-implant region, 45, 58, 59
Brain
  computer interface (BCI), 240, 244, 254
  to-brain system (BTBS), 243, 244, 254

C

Calibration, 61, 62
Cancellous bone, 47, 49, 50
Cancer detection, 128, 130
Cardiovascular disease (CVD), 224, 236
Central composite design (CCD), 108
Cesium iodide (CsI), 122, 123, 134
Chatbot, 22, 25, 28–32, 34, 41, 42
Chemical composition, 102, 105

Circuit, 2–4, 6, 7, 9, 12, 28, 162, 170, 208, 213, 219, 244
Classification, 23, 24, 118, 189, 223, 225, 235
Cochlear implant (CI), 178, 241, 254
Coefficient, 6, 64, 116, 142, 174, 190
Common carotid artery (CCA), 224
Components, 3, 9, 26, 27, 34, 40, 75, 77, 78, 83, 86, 95, 96, 98–100, 105, 109–111, 120, 127, 144, 158, 193, 195, 224, 248, 251, 268
Computed tomography (CT), 67, 123, 124, 127, 133
  scan, 67, 124
Computer
  generated (CG), 67
  vision, 62, 63, 115, 129–132, 134, 260, 262
    techniques, 130
    technologies, 132
Computerized information systems (CIS), 245
Continuous positive airway pressure (CPAP), 209, 221
Contour
  number, 107
  width, 107
Conventional
  mode, 75
  production methods, 100
Conversion, 120, 122, 123
Convolutional neural networks, 24, 43
Coronavirus disease (COVID-19), 128, 132, 201, 207, 208, 210, 221, 263, 273
Cortical
  bone properties, 50
  region, 51
Covariance-Based SEM (CB-SEM), 250, 251
Critic
  approach, 169, 180
  method, 183
Cubage, 190
Cues, 65, 247
Cumulative distribution function, 191
Custom rehabilitation devices, 143
Cyanoacrylate cement, 51
Cyber
  medicines, 245
  narcotics, 245
Cyborg, 239–241, 243–245, 249, 250, 254
  operations, 239
  radio waves, 244
  service, 239
  surgeon, 239
  technologies, 241

D

Data
  analysis process, 250
  envelopment analysis (DEA), 171, 184
  patterns, 246
  structures, 24
Database, 25, 166, 223, 227, 231, 235
Decision-making, 169, 171, 180, 201, 240
Deep
  brain stimulation (DBS), 241, 242, 254
  deterministic policy gradient (DPG), 248
  learning systems, 24, 62, 78, 130–132
  Q network (DQN), 248
Degrees of freedom (DOF), 25, 138, 140, 145, 146
Density, 23, 40, 49, 85, 105–107, 109, 110, 203
Dental
  implant, 45–50, 59
  implantology, 51
  prosthesis, 46, 47
Dentistry, 46, 47, 51, 143
Depth
  estimation, 61, 63–68, 72
  from the defocus (DFD), 68
  map evaluation, 67
  prediction, 65
Design data book, 33
Developed library management, 35
Diagnosis, 116, 120, 127–132, 134, 224, 247, 263
Diagnostic, 116, 201, 258
  information, 115
  tasks, 130
  techniques, 273
Dialog, 22, 28, 29, 34
  skill, 22, 28, 29
Diameter, 45, 47, 50, 52–54, 56, 58, 59, 81, 87, 88, 97, 162
Dielectric
  elastomer
    actuators, 143
    substrate (DES), 142, 146
  layer, 141

Digital
format, 116, 244
images, 116
signals, 48
Dilation operation, 120
Dimensions, 10, 47, 50, 62, 98, 254
Diodes, 2, 4, 9
Direct
conversion, 123
current (DC), 3, 7, 12, 27, 77, 137, 140, 144–146, 158, 159, 210, 213, 215, 220, 229
execution, 23
metal laser sintering (DMLS), 96, 100
motor, 26
Disinfection systems, 260
Disparity, 63–69, 72
map, 64, 68, 69, 72
Distal side, 45, 51, 58, 59
Distortion, 61–63, 100
Double-screw-drive (DSD), 138, 139, 146
Drone, 75–88, 91, 187, 188, 193, 196, 199–203
battery, 85
monitoring, 202
systems, 188, 193, 199–202
Drug
delivery systems, 273
resistant, 272
Duodenum, 67
Dystonia, 241

E

Edge, 25, 62, 116, 119
based technique, 119
detection algorithm, 119
Elastic
fluidic actuators, 140, 146
inflatable actuators, 140
Elbow orthosis, 144
Electric
polarization, 126
voltage, 126
Electrical
circuit, 28
components, 27
energy, 26, 126, 137
fields, 126
resistance, 47, 48
rotary actuator, 138
signal, 26, 48, 157, 224, 245
strain gauge, 48
Electroactive polymer, 141, 143
Electrocardiogram (ECG), 219–221, 223–227, 231–235
Electrodes, 123, 126, 142, 241, 244, 266
Electroencephalography (EEG), 152, 243, 254
Electromagnetic
induction, 27
radiation, 121
Electromechanical, 26
Electromyographic, 144
Electron optic system, 121
Electronic
circuits, 34
device, 26
equipment, 27
speed controller (ESC), 85, 86
Electrons, 121, 123
Electrothermal actuation, 142
Embedded robot, 37
Emerging, 2, 18, 126, 224
Emotional status, 247
Emphasis, 192, 239
Encode, 244
Endeavor, 21
Endoscopy, 65–67
camera, 65
images, 65
Endotracheal
intubation, 209
tubes, 207
Endurance, 170, 200, 240
Energy harvesting, 1, 18
Enhanced, 122, 132, 211, 252, 272
permeability and retention effect (EPR), 272
Entropy
approach, 169
method, 183
Environmental welfare, 86, 87
Epilepsy detection system, 152
Epipolar line, 71, 73
Epoxy sheet, 50
Equal weightage, 183

approach, 169, 174
method, 169
Erosion operation, 120
Essential tremor, 241
Evolutionary robotics (ER), 249
Extension, 48, 67, 145, 241
Extensometer, 48
Extrinsic parameters, 62

F

Fabrication, 34, 98, 107, 269
Fatigue, 47, 102, 105, 116, 134
Feature extraction, 120, 226, 231
Federal Aviation Administration (FAA), 201
Field
innovation team (FIT), 201
programmable gate array, 27
Filament, 81, 97, 109
Finite element analysis (FEA), 46, 59, 108
First
generation ventilators, 208
person view (FPV), 86
Flat panel detector (FPD), 122, 123, 134
Foams, 49
Forceps, 138–140
Frame, 2, 27, 77, 78, 83, 85, 86, 107, 125, 193–195, 211
Frequency, 2, 4, 6, 7, 9, 15–17, 78, 126, 141, 226, 231, 232
Friction, 138, 140, 144, 145
Fundamental block diagram, 64
Fused
deposition
method, 28
modeling (FDM), 81, 92, 95–99, 107–111
filament fabrication (FFF), 97
Fuzzy
axiomatic design (FAD), 171, 184
logic (FL), 110

G

Gadgets, 2, 202, 203
Gadolinium element, 127
Gallium
arsenide (GaAs), 4, 18
nitride (GaN), 1–6, 9, 12–18
circuit model, 6
rectifier circuits, 1, 2, 18
schottky diode, 6, 14, 16
Gamma, 116, 125
Gantry, 123, 124
Gastro endoscopy procedure, 67
Gauge, 48, 51, 133
Gauss distribution curve, 191
Gaussian distribution, 190
Generalized seizures, 154
Genetic
algorithm (GA), 105–107, 110, 226, 249
engineering, 240
Geographical barriers, 75
Geometric
clues, 62
design, 140
objects, 120
relation, 63
Geometry, 47, 63, 100, 133
Georgia Tech Research Institute (GTRI), 190
Germanium rectifier, 3
Gimbals mechanism, 138
Glass
capsule, 244
filled nylon, 100
Global
positioning system (GPS), 25, 78, 92, 187, 199, 200, 203
radiometric variation, 70
Gradient image, 25
Graphical programming, 35, 38–40
Grave consequences, 241
Gray
level, 121
value, 118, 119
relational analysis, 170, 184
Guidance, 35, 192
Guidelines, 51

H

Haptic
actuator, 141
sensations, 141
technology, 141
Hardware, 35, 63, 68, 97, 155, 159, 212
Healthcare
applications, 200, 258

area, 115
computer vision systems, 131
delivery, 75
operations, 189
professionals, 259, 260
sector, 188, 199, 202, 247
service, 252, 254
sector, 239
systems, 187
Heart
disease, 224, 264, 273
rate, 161–163, 216, 217, 231
rhythmic disorders, 2
Heat conduction, 101
Hemoglobin, 161
Heterotrait-monotrait ratio (HTMT), 251, 252
Hexacopter, 76
High
frequency
alternating voltage, 126
sound waves, 126
power
diodes, 4
laser beam, 101
resolution cross-sectional images, 128
Homogeneity property, 119
Hospital-acquired infection, 274
Hot spots, 125
Hovering, 188, 189, 192
Human
anatomical body, 115, 134
assistance, 22, 39, 42
brain, 24, 243, 244
computer interfaces, 61, 72
immunodeficiency virus (HIV), 200, 201
individual, 25
intervention, 78, 254
machine interaction, 243
perception, 141
services, 239, 250, 254
surgeon, 239, 240
vision, 61
Hybrid
model, 67
segmentation algorithmic program, 67
Hydraulic energy, 143
Hydrogel material, 140

I

Identification, 118, 120, 124, 130, 132, 231, 271, 274
Ideology, 166
Image
acquisition, 67, 117
detection, 24
enhancement, 118
intensifier system, 121, 122
interpretation system, 121
processing, 24, 25, 35, 63, 115–119, 121, 129, 130, 134
technique, 25
segmentation, 118, 119
stabilization, 68
Imaging system, 116, 121, 122
Immobile patients, 137
Immune system, 138
Impedance, 1, 4, 7–9, 11, 18, 140
matching
circuit, 4, 8, 9, 11
network, 7
Implant, 45–50, 52, 53, 58, 59, 105, 240, 241, 244, 245, 264
biomechanical behavior, 46
bone interface, 46
dentistry, 48
diameter, 45, 47, 58, 59
longevity, 46
survival life, 46
Implantation, 46, 241, 245
In vitro, 47
In vivo, 47, 48
Incorporated carbon nanotube position (ICNP), 142
Indicator, 34, 35, 122, 251
Indigo carmine (IC), 67, 215, 217
Indoor, 78, 162
Induced, 46–48, 58, 142
Industrial, scientific, and medical (ISM), 2
Inflatable structures, 143
Infrared light source, 161
Input
layer, 122
phosphor, 121
power, 1, 2, 15–18
Instrumentation program, 35
Integration, 63, 130, 137, 146, 258
system, 63

Intelligence, 22, 42, 145, 188, 203, 240, 246, 247
surveillance, target acquisition, and reconnaissance (ISTAR), 188
Intensive care units (ICU), 208, 212, 220, 221
Intent, 22, 28, 29, 34, 119
detection, 30
Intermittent ventilation, 208
Internal, 27, 62, 100, 116, 117, 121, 133, 215, 251, 261, 274
anatomical structure, 120
threads, 27
Internet of things (IOT), 77, 78, 137, 146, 151, 155, 156, 166, 203, 258, 274
Invasive, 67, 116, 138, 146, 209, 221, 244, 269
ventilation, 209, 221
Investigation, 59, 105, 107–111, 244
Ionic
conductive layers, 141
strength, 140
Ionizing radiation, 126, 127
Irrelevance detection, 30
Isotropic material, 97

J

Jaw, 46, 140
Jig fixtures, 51
Joining process, 27
Jumper wire, 28
Junction, 4, 6, 46
capacitance, 4, 6

K

Kalma filter, 25
Kinematic performance, 140
Kinetic energy, 137

L

Laboratory
robots, 266
virtual instruments engineering work-bench, 35
LabVIEW software, 27, 34, 35, 38, 39
Laminated bone blocks, 49
Lane edge, 25
Laparoscopic
forceps, 140
grasper, 140
surgical navigation, 64
Laser melting, 101, 102
Lean management, 129
Library
administration system, 25
automation, 25, 42
issuance system, 25
management system, 22, 31, 34, 35, 37–39, 41, 42
Ligaments, 127
Light
detection and varying (LiDAR), 262
photons, 121
weight, 49
Limbs, 266, 267
Limit reduction strategy (LRS), 229
Line follower chassis, 27
Linear, 62, 138, 140, 144, 145, 171, 214, 264
Lithium polymer (LiPo), 85, 92
Load
resistance, 15, 16
transferring mechanism, 46
Local gradient information, 25
Logic signals, 116
Low
carbon steel powder, 106
loading applications, 51
resources, 115
voltage drop diode, 3
Lower jaw bones, 46
Lymphatic system, 273

M

Machine
learning (ML), 23, 24, 43, 130, 226, 246, 247, 258, 274
methods, 247
operations, 23
technologies, 246
vision, 35, 62, 129, 130
Magnetic
actuators, 140
field, 127
resonance imaging (MRI), 127, 128, 130, 133
scan system, 127

Magnitudes, 46
Mail intimation, 35
Malignancies, 127, 269, 272
Mandible, 46, 50
Manipulation, 24, 246, 247, 260, 269
Manual assistance, 155
Markov random field (MRF), 65
Mask-R convolutional neural networks (Mask R-CNN), 130
Massive length fabrics, 271
Matching network, 3, 6, 7
Material, 34, 47, 48, 75, 77, 83, 98, 100, 109–111, 126, 127, 269
  properties, 49, 50
  science, 244
  surfaces, 48
Mathematical functions, 116
Matrix, 24, 69, 120, 123, 172, 173, 175, 178, 250
Mechanical, 21, 26, 42, 48, 49, 105–111, 126, 137, 140, 143, 144, 207–212, 215, 219, 220, 268
  department rack, 42
  energy, 26, 126
  intricacies, 46
  pressure, 126
  quantities, 48
  rack, 21, 30, 41, 42
  ventilator, 207, 210, 220
Mechatronics
  instruments, 31
  system, 25
Medical
  aids, 75
  applications, 62, 189, 199, 200, 226
  care, 77, 115, 208
  devices, 137
  equipment, 75, 77, 201, 202
  errors, 115
  image, 115–117
    interpretation system, 121
  imaging
    applications, 72, 73, 120
    systems, 121
    techniques, 128
  products, 75
  robots, 258, 259, 263, 266
  training, 128, 131
  ultrasound frequencies, 126
Medication, 115, 134, 262, 271–273
Melanomas, 130
Memory
  capacity, 170, 172, 181, 183
  space, 172
Meningitis, 153
Merge, 66, 119, 152
Mesial, 45, 51, 58
Metabolic action, 125
Metal
  electrode, 4
  powder, 101
Methodology, 4, 31, 58, 59, 66, 96, 97, 163, 218, 231, 232
Micro strips, 1
Micro-actuators, 137, 138
Micro-device, 137
Micro-fabrication sciences, 244
Microfluidic actuators, 138
Micrograsping forceps manipulator, 138
Micrometers, 137, 269
Micromotion, 137
Micro-motor, 140, 146
Microprocessor, 27, 208, 226
Micro-rubber pneumatic actuators, 140
Microsoft access, 25
Micro-strain, 47
Microstrip, 5–7, 9, 10, 12
  dimensions, 6, 9, 10
  length, 6
  matching circuit, 5, 6
Microwaves, 121
Miniature tool, 138
Minimal invasive surgery (MIS), 138, 146
Minimally invasive surgery (MIS), 67, 137–141, 146
  application, 67, 140
  workspace, 140
Minor quantity, 125
Mitigate, 207
Modular robots, 262
Molten powder, 101
Monitoring, 128, 129, 132, 152, 163, 202, 209, 225, 258, 259, 262, 269
Monocular, 64–66
  endoscopic videos, 66
Monotonous duties, 266
Morphological
  methods, 119

operations, 119
Motion
control, 26, 34, 246
kit, 26
sensor, 152
Motor
drive circuit, 25
driver, 27, 215, 216
Multi-criteria decision-making (MCDM), 170–172, 184
Multidirectional bending, 138
Multi-DOF rotation & ultrasonic actuator, 140
Multimodal seizure recognition systems, 152
Multiple reproduction, 116
Multi-ring detectors, 125
Multirotor, 188, 196, 197, 199
drones, 76, 196
Multi-view
stereo method, 66
system, 67
Myocardial infraction (MI), 235
MyRIO hardware, 27, 34

N

Nanoactuators, 269
Nanoelectromechanical systems (NEMS), 269
Nanofabrication techniques, 269
Nanoimprint lithography, 273
Nanomedicine, 257, 258, 273, 274
Nanometer, 269
Nanomotors, 269
Nanoparticles, 269, 270, 273, 274
Nanorobotics, 268, 269
Nanoscale, 257, 269, 271–274
Nanoscience, 268
Nanoseconds, 4
Nanosensors, 269
Nanotechnology, 268–273
drug delivery, 272
National
Health Laboratory Service (NHLS), 201
Instruments (NI), 35
Navigation, 37, 41, 61, 64, 68, 72, 189, 192, 193, 199, 264
Negative-pressure ventilation, 208
Nerve fibers, 244
Nervous system, 241, 268
Neural networks, 24, 258
Neuro feedback, 243
Neuromotor prostheses (NMPS), 241, 242, 254
Neuromuscular machine, 266
Neuromusculoskeletal prosthesis, 266
Neurons, 24, 241
Neuropsychiatric problems, 241
Neuroscientists, 244
Neurostimulator, 241
Noisy sensor measurements, 192
Non-destructive methodology, 58
Non-invasive
features, 209
mechanical ventilators, 209
ventilation, 209
Nonlinear processes, 62
Non-melted powder, 101
Normal sinus rhythm (NSR), 235
Normalized cross-correlation (NCC), 61, 71, 72
Nozzle, 81, 97
N-type semiconductor, 4
Nuclear medicine, 125
Numerical algorithms, 24
Numerous-orienting systems, 65
NumPy, 24
Nylon, 95, 98, 100, 109

O

Object
boundary, 120
recognition, 25, 66
Observation, 23
Obsessive-compulsive disorder, 241
Obstacle detection, 26
Octocopter, 76
Odometry, 25
Online chat, 22, 28
Open porosity, 49
Operating
aerial systems, 188
frequency, 3, 4, 12
Operational challenges, 190
Optical, 48, 62, 71, 87, 129, 156, 269
coherence tomography (OCT), 128, 129
measures, 128
scan, 128

sensors, 161
strain gauges, 48
Optimal
configuration mechanism, 140
specifications, 34
Optimized microstrip dimensions, 10
Optimum angulation, 140
Optoelectronic strategy, 4
Ordinary least squares (OLS), 250
Organic
cells, 269
techniques, 257
Organizational symbolism, 245
Orientation, 62, 65, 88, 102, 106–108, 110, 228
Orthopedic surgeries, 261
Oscillator, 2, 213
Osseo, 46
Osseointegration state, 47
Output
phosphor, 122
voltage, 6, 7, 10, 12, 13, 229

P

Pacemaker, 1, 2, 18
Pandas, 24
Parameter, 6, 40, 58, 62, 63, 68, 96, 98, 105–111, 151, 217, 226, 228, 250
tuning process, 62
Parkinson's disease, 241
Partial least squares SEM (PLS-SEM), 239, 250, 251, 254
Particle swarm optimization (PSO), 110, 226–228, 230, 235, 236
Pattern calibration, 62
Penetrate, 126
Perception, 65, 106, 141, 240, 247
Peripherals, 155
Permeability, 272, 273
Petition phases, 67
Pharmacology, 240
Phosphorus substance, 121
Photocathode, 121
Photodiode, 161
Photoelasticity, 46
Photoelectric effect, 121
Photons, 121–123
Physical
attributes, 240
capabilities enhancement, 240
Physiological cues, 247
Picoseconds, 4
Piezoelectric
actuators, 137, 146
effect, 126
material, 126
phenomenon, 126
Piston-cylinder, 137, 140, 146
solutions, 140
Pivoting
mechanism, 138
motion, 138
Pixel, 62–65, 68, 69, 71, 72, 118–120, 123
block, 68, 69
Plotting 2D, 24
Pneumatic
actuation, 143
actuators, 140
artificial muscles, 143
concepts, 33
Pollution treatments, 272
Polycarbonate plastics, 100
Polyether ether ketone (PEEK), 95, 108–111
Polyethylene terephthalate glycol (PETG), 28, 43, 95, 98, 107, 109, 111
Polylactic acid (PLA), 28, 40, 77, 95, 98, 107–109, 111
Polymeric material, 107
Polymerization process, 49, 97
Polyurethane, 48–50, 59, 141
foam, 48–50, 59
Porosity, 105, 106
Portable microstimulator, 243
Positive
ideal solution (PIS), 173, 174, 176–179, 181, 182
predictivity (PP), 226, 232
pressure ventilation, 209
Positron emission tomography (PET), 125, 134
Post-operation, 138
Postoperative pain, 138
Potentiometer, 213, 215, 216, 219
Power
amplifier, 2
density, 85, 105
electronics, 4
Precision, 66, 77, 95, 143, 170, 240, 260, 264, 271

Premature, 125, 128, 130, 134
ventricular contraction (PVC), 232
Primary goal, 21, 224, 252
Printing technique, 28, 97, 108
Propeller, 76, 77, 83, 86–88, 190, 196, 199, 202
Proportional-derivative (PD), 192, 203
Prosthetic
limb, 266, 268
parts, 48
Prosthodontic treatment, 46, 59
Protocol, 51, 78, 212, 218
Protons, 127
Prototype, 78, 132, 156, 166
spectacles, 157
Prototyping tools, 28
Pulse oximetry, 216, 217
Python, 24
language, 61

Q

Quad-band GSM, 160
Quadcopter, 76, 78
configuration, 199
Qualitative approach, 47
Quality assessment, 116
Quantitative
approach, 47
evaluation, 265

R

Radiant energy, 121
Radio
frequency (RF), 2–4, 18, 78
waves, 121, 127
Radioactive chemicals, 125
Radiography, 122, 132
Radiotherapy, 264
Radiotracer, 125
emitting energy, 125
Rapid
advancement, 75
restoration function, 217
Raster
angle, 107–109
width, 107
Ratio measures, 68
Receiver, 2, 3, 77, 85
Recommendation, 23
Reconnaissance, 188, 203
Rectangular rosette strain gauge, 51
Rectenna, 2, 3
Rectification, 4, 6, 190
Rectifier, 1–4, 6–10, 12–18
circuit topologies, 2, 6
efficiency, 3, 9, 16
Recurrent neural networks, 24, 43
Region-based technique, 118, 119
Regression, 23
analysis, 106
Rehabilitation, 132, 143, 144, 146, 264, 265
devices, 137, 143, 146
robots, 264, 265
Reinforcement
learning, 247, 248
mechanism, 248
Relevant works, 25
Remote transmission, 17
Rephrase, 29
Rescue, 188
Reset button, 213, 214
Resistance, 1, 2, 6, 9, 18, 48, 49, 105, 157, 158, 161, 215
Respiratory apparatus, 207
Retention effect, 272, 273
Retina, 128
Retinal
edema, 128
stimulation, 244
Return loss, 1, 7, 9, 10, 18
Reynolds, 190
Rigid polyurethane structure, 49
Robot, 21, 22, 25–27, 31, 34, 37, 40–43, 61, 72, 130, 134, 142, 144, 146, 169–172, 183, 184, 239, 240–247, 249–251, 254, 261, 263–268
application, 170
assisted surgery, 263
development, 25
surgeon, 130, 134, 239
technologies, 245
Robotic, 1, 21, 27, 42, 45, 61, 65, 66, 75, 95, 115, 137, 142, 151, 169, 187, 207, 221, 223, 239, 249, 257, 258, 261, 262, 264–266, 268

arm, 21, 25, 43, 210, 211, 219, 220, 241, 262
hysterectomy, 261
laws, 21
library system, 25
prostatectomy, 261
prosthetics, 266
technology, 239, 252
therapy, 262
vision, 65
Robust lane detection methods, 25
Rotary motor, 26
Rough depth measurements, 63

S

Salient features, 28
Scanning
electron microscopy–secondary electron (SEM-SE), 105
technologies, 127
Schematic diagram, 100, 101, 271
Schottky diodes, 1, 4, 18
Science fiction novels, 245
Scientific computations, 24
Scikit learn, 24
SciPy, 24
Seizure, 151–156, 159, 163, 166, 243
detection, 152
symptoms, 166
Selective laser
melting (SLM), 95, 96, 101, 102, 105, 109, 111
process, 101, 102
sintering (SLS), 95, 100, 106
process, 100, 101, 105
technique, 101
Self-esteem, 46
Selfsupervised method, 66
Semi global (SG), 48, 51, 58
block
algorithm, 61
matching (SGBM), 61, 64, 70, 72, 73
simulation (SGM), 70, 71
Sensitivity (Se), 152, 171, 232, 234
Sensor, 26, 48, 64, 78, 87, 88, 124, 127, 145, 151, 155–157, 159, 163, 189–191, 217, 243, 249, 258, 268
device, 47
flashes, 161
measurements, 190–192, 199
methodology, 58
Sensory processing, 240
Series
diode, 6, 7, 9, 12, 15, 16
half wave rectifier, 12
rectifier, 6
half-wave gan rectifier, 7, 12
Servo motor, 27, 210, 212–215, 219
Shape memory alloys (SMA), 137, 140, 144, 146
actuator, 140
Short
term fourier transform (STFT), 227, 236
tidal volume, 211, 221
Shrink, 126, 140
Shunt, 3, 4, 6, 7, 9, 10, 13–16
diode rectifier, 6, 7, 15
half wave gan rectifier, 9, 13
Silicon
carbide (SiC), 4, 5, 18
chip, 244
devices, 4
diode, 4
rectifier, 3
Simulation results, 1, 12, 15, 18, 235
Single
photon emission computed tomography (SPECT), 125, 134
rotor helicopter, 76
Sintered products, 106
Skew, 62
Skin disorders, 130
Slited hole, 138
Slope limits, 61
Small
electrical sensor, 48
signal diodes, 4
size template, 120
Smart
inbuilt techniques, 21
intelligence system, 21
tissue autonomous robot (STAR), 130
Soft robotic actuators, 140
Software application, 22
Solid rigid polyurethane foam, 49, 50
Sonar data, 25

Spatial
frequency, 120
resolution, 122
Specimens, 52, 106, 109
Speed controllers (SCs), 77, 92
Spherical rotor, 140
Spinal cord, 127, 130
Stainless steel, 95, 101, 105, 109
Statins, 273
Statistical
design, 106
models, 170
Steady-state visual evoked potential (SSVEP), 244
Stem cell injury, 271
Stereo
communication algorithms, 61
cues, 65
depth measurement algorithm, 65
imaging, 63, 67
lithography (SL), 33, 40, 95, 111, 198
matching, 64, 67
photogrammetry, 48
signals, 65
vision, 61, 63–65
Stereolithography (STL), 81, 95–98, 100, 106, 133
Strain
dispersion, 45, 59
distribution characteristics, 46
gauge, 47, 48, 51–53, 58, 59
method, 47
measuring techniques, 48
value, 47, 48, 58
Stress dispersion, 46
Structural equation modeling (SEM), 250
Structuring element, 120
Subfield, 23, 24
Sub-millimeter residual error, 67
Sum of absolute difference (SAD), 61, 64, 68, 69
Supervised learning, 23, 66, 258
Surgeons, 65, 66, 130, 131, 133, 209, 239, 240, 254, 260, 262
Surgery, 2, 66, 67, 129–133, 138, 141, 207, 209, 210, 239, 261, 264, 268
Surgical
assistance robots, 260
robotic, 264
robots, 260–262
services, 240, 254
Surveillance, 46, 62, 188, 202, 203
Symptoms, 130, 154, 211, 226

T

Tangential, 195
Tartar, 137, 143, 146
Technical
call, 48
problems, 47
Technique, 48, 95–98, 100, 101, 105, 107, 108, 116, 118, 119, 125, 127, 169, 171, 207, 226, 241, 250, 251, 264, 268, 269, 271
for order of preference by similarity to ideal solution (TOPSIS), 169, 171–173, 183
Telemedicine, 200
Telepresence, 262
Telerobotic system, 140
Telescopic cylinder, 21, 43
Temperature sensor, 151, 156, 158, 166
Temporary joining process, 27
Tendon, 127, 245
driven, 137, 139, 146
Tensile, 48, 102, 105–109
properties, 105, 107, 108
Test kits, 75
Therapeutic intervention, 240
Therapist, 265
Therapy, 120, 121, 132, 211, 241, 262–264, 271, 273
Thermodynamics, 33
Thermoplastic filaments, 97
Thermostat
devices, 158
module, 158
temperature controller module, 166
Thin
beams, 123
film transistor (TFT), 123, 134
Three-dimensional (3D), 24, 28, 34, 40, 61–63, 65–67, 72, 73, 75–81, 83, 86, 87, 91, 95–100, 102, 105–111, 120, 123–127, 133, 143, 261, 262, 274
bio-printing, 133
geometric model, 120

image, 63, 123, 124
model, 28, 67, 133
object, 133
printer, 28, 40, 77, 81, 98
printing, 28, 34, 75–77, 81, 95–97, 102, 105, 107–111, 133
technique, 28
technology, 77, 96, 133
visual assist surgeons, 65
Thresholding, 25, 118
approach, 118
strategy, 25
Thrust force, 87
Tissues, 124–127, 138, 143, 145, 224, 261, 262
Titanium, 101
Tomography, 116, 118, 128
Topologies, 4, 6
Trachea, 207, 210
Tracheostomy, 207, 209
Traditional algorithm, 68
Traffic scenario, 25
Trajectory, 193, 195
fixed axes, 193
Transducers, 126
Transfer connection blocks (TCBs), 66
Transhumanism, 240
Transient simulation, 10, 13
waveforms, 15
Transmitter, 2, 77, 78, 85, 152
Transmitting antenna, 2
Tricopter, 76
Tubular construction, 122
Tumor, 121, 130, 131, 143, 272
detection, 128, 130, 131
tissue, 121, 272
Two-dimensional (2D)
image
object, 62
planes, 63
matrix images, 24
Typical matching impedance, 4

U

Ultimate model, 49
Ultrasonic
sensor, 26
sound waves, 26
waves, 126
Ultrasound, 87, 117, 125–128, 224
transducer, 126
Unconventional missions, 189
Unipolar devices, 4
Universal testing machine (UTM), 51, 59
Unlabeled
data, 23
training data, 67
Unmanned aerial vehicle (UAV), 63, 75, 92, 187–189, 192, 201–203
Unorthodox technology, 190
Unrestrained electrical disorder, 153
Unsupervised learning, 23, 258
Urological surgeries, 67

V

Vaccines, 75, 201
Vacuum tubes, 121
Vagus nerve stimulation (VNS), 243, 254
Vanishing points, 62
Vector, 193, 197, 227
notation, 193
Vehicle-fixed axes, 193
Velocity, 87, 144, 170, 192, 194, 195, 228, 229
Ventilation device, 208
Ventilator, 207–212, 215, 219–221
Ventricular
fibrillation (VF), 235
flutter (VFLU), 235
tachycardia (VT), 211, 235
Vertex, 120
Vertical take-off and landing (VTOL), 76, 188, 196, 197
Vibration amplitude, 141
Victimization method, 67
Video analysis, 130
Virtual
environment, 132
reality (VR), 132, 134
Visible light, 121, 122
Vision restoring technology, 244
Visual
grasping, 25
line-of-sight (VLOS), 201
path, 244
representation, 118
sensor, 62

Visualize, 125, 270
Voice-based system, 22
Voltage doubler, 1–4, 6, 7, 9, 14–16, 18
gan rectifier, 9, 14
rectifier, 1–3, 6, 7, 9, 15, 16, 18

W

Waterborne polyurethane (WPU), 142, 146
Watertight parts, 97
Watson
assistant, 22, 25, 28–30, 34
chatbot, 22, 25, 31
tool, 29
model, 22, 29
Wavelet transform (WT), 223, 226, 231, 232, 234–236
Wearables technology, 240
Webhooks, 30
Website, 23, 29
Weightages, 105, 169
White-light endoscopic images, 65
Wide range, 1, 2, 15, 18, 35, 101, 120, 127, 188, 269
Wireless
communication channel, 244
power transmission (WPT), 1–3, 18
applications, 4
system, 1–3, 18
Workflow of chatbot, 30
Working principle, 86, 124, 128
Workplace environment, 259
Workshop technology, 33
Workspace, 138, 140
World
coordination, 25
Health Organization (WHO), 224, 236, 258
Worldwide, 75, 188, 224
Wound treatment, 274
Wraps, 244
Wrist bracelet, 152

X

X-ray, 116, 121–124, 127, 128
films, 121
image intensifiers, 121
imaging systems, 121
output, 122
photons, 121
sensors, 124
tube, 123

Y

Yaw motions, 199
Young's modulus, 107

Z

Zero, 88, 190, 228, 231
Zirconium titanate, 127
Zones, 125